T0374617

Philosophy of Physics

PRINCETON FOUNDATIONS OF CONTEMPORARY PHILOSOPHY

Scott Soames, ***Series Editor***

Philosophical Logic, John P. Burgess
Philosophy of Language, Scott Soames
Philosophy of Law, Andrei Marmor
Truth, Alexis G. Burgess and John P. Burgess
Philosophy of Physics: Space and Time, Tim Maudlin
Philosophy of Biology, Peter Godfrey-Smith
Epistemology, Ernest Sosa
Philosophy of Mathematics, Øystein Linnebo

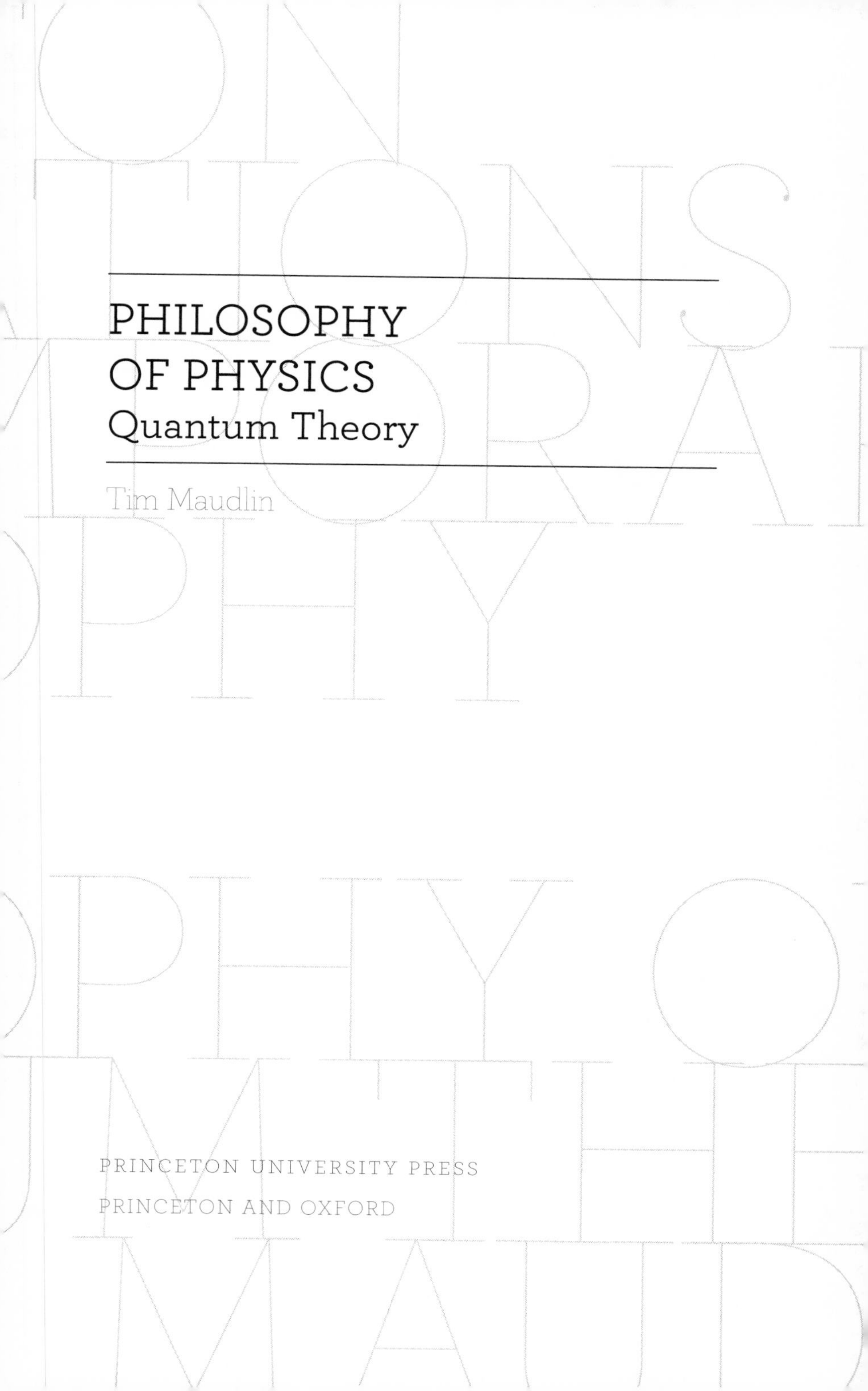

PHILOSOPHY OF PHYSICS
Quantum Theory

Tim Maudlin

PRINCETON UNIVERSITY PRESS
PRINCETON AND OXFORD

Published by Princeton University Press
41 William Street, Princeton, New Jersey 08540
6 Oxford Street, Woodstock, Oxfordshire OX20 1TR

press.princeton.edu

Library of Congress Control Number: 2018949371
ISBN 978-0-691-18352-7

British Library Cataloging-in-Publication Data is available

Editorial: Rob Tempio and Matt Rohal
Production Editorial: Leslie Grundfest
Production: Jacqueline Poirier
Publicity: Jodi Price
Copyeditor: Cyd Westmoreland

This book has been composed in Archer and Minion

Printed on acid-free paper. ∞

Printed in the United States of America

10 9 8 7 6 5 4 3

To Shelly Goldstein: physicist, philosopher, mathematician, dear friend

Contents

Introduction

THIS VOLUME OF *Philosophy of Physics* confronts quantum theory. The original intent was to cover both quantum theory and statistical explanation, but that was not feasible, given the constraints of space. Quantum theory presents a fiendish challenge for a book like this: There are too many phenomena, too much technical elaboration, and too many fundamental conceptual issues to be adequately exposited in such a limited span. Unlike space-time theory, where there is substantial agreement about how to understand the best physics we have (General Relativity), quantum theory has always been a battleground of contention. Nothing one can say would command the assent of most physicists or philosophers.

Structuring the manuscript demanded painful choices about what to present, the appropriate level of technical complication, what historical background to include, which controversies to mention, which alternative elaborations of theories to consider. Every decision was difficult and can be legitimately challenged. Important phenomena and theoretical approaches have been left unmentioned. Ideas for reconciling quantum theory and General Relativity—quantum theories of gravity—are not discussed. All but the last chapter deal solely with nonrelativistic quantum theory.

What principle guided these choices? The central problem facing attempts to understand a quantum theory is how it manages to model empirical phenomena in a principled way. This is often referred to as "the measurement problem," because the sorts of laboratory operations used to provide data are called "measurements." But the problem has a much wider scope. Any macroscopic phenomenon can in principle test a fundamental physical theory, because the theory should be able to provide a physical account of it. Erwin Schrödinger famously asked how quantum theory could model how a cat in a particular experimental setting ends up either alive or dead. It is irrelevant for his point whether the experiment counts as a "measurement."

John Stewart Bell made a proposal about how this can be done, which he called the *theory of local beables*. "Beables" refers to the ontology of a theory: what it postulates to exist. "Local" indicates a beable that exists in a small region of space or space-time. Fixing the distribution of local beables at a microscopic scale fixes the location, shape, and motion of their macroscopic aggregates and thereby can solve the measurement problem and Schrödinger's cat problem. What one needs from such a theory is an inventory of local beables and an account of their dynamics: how they get distributed in space-time.

This basic idea can be implemented in different ways, which can be illustrated in a nonrelativistic setting where the technical details are easier to grasp. These are admittedly empirically inadequate theories, but they provide models of general strategies for solving the measurement problem. They also illustrate many iconic quantum-mechanical effects. The additional challenges facing relativistic extensions can be considered later. So our investigation proceeds by discussing three ways to implement this strategy nonrelativistically, together with a short discussion of the additional challenges facing extensions to a relativistic space-time.

This approach faces perils. If the correct solution to the measurement problem does not involve local beables, or if those beables have no nonrelativistic analogs, then starting with nonrelativistic quantum mechanics is counterproductive. But one has to start somewhere, and in an introduction, it is best to start with what is easiest to grasp. If nothing else, nonrelativistic quantum mechanics can act as a foil for alternative theories, so one can see how the assumptions made here fail. Starting from what we understand and seeing clearly its inadequacies can provide a path to conceptual progress.

By far the most controversial aspect of this book is not what it contains but what it omits. There is detailed discussion of the Ghirardi-Rimini-Weber spontaneous collapse theory, of the pilot wave theory of the Louis DeBroglie and David Bohm, and of Hugh Everett's Many Worlds theory. But there is no discussion—indeed aside from here no mention—of the most famous "interpretation" of quantum theory of all: the Copenhagen Interpretation ascribed to Niels Bohr and his colleagues. Why is that?

A physical theory should clearly and forthrightly address two fundamental questions: what there is, and what it does. The answer to the first question is provided by the *ontology* of the theory, and the answer to the second by its *dynamics*. The ontology should have a sharp mathematical description, and the dynamics should be implemented by precise equations describing how the ontology will, or might, evolve. All three of the theories we will examine meet these demands.

The Copenhagen Interpretation, in contrast, does not. There is little agreement about just what this approach to quantum theory postulates to actually exist or how the dynamics can be unambiguously formulated. Nowadays, the term is often used as shorthand for a general instrumentalism that treats the mathematical apparatus of the theory as merely a predictive device, uncommitted to any ontology or dynamics at all. That predictive device is described in Chapter 2 under the moniker "the quantum recipe." Sometimes, accepting the Copenhagen Interpretation is understood as the decision simply to use the quantum recipe without further question: Shut up and calculate. Such an attitude rejects the aspiration to provide a physical theory, as defined above, at all. Hence it is not even in the running for a description of the physical world and what it does. More specific criticisms could be raised against this legacy of Bohr, but our time is better spent presenting what is clear than decrying what is obscure.[1]

Besides rejecting the usual terminology of "quantum theory" versus "interpretation of quantum theory" in favor of "predictive recipe" versus "physical theory," and besides ignoring the historical question of what (if anything) should count as the Copenhagen Interpretation, this book differs from most standard discussions in a third way. It has become almost de rigueur in the quantum foundations literature to systematically misuse the terms "realist,"

[1] More details about the obscurity can be found in Norsen (2017), Chapter 6, and throughout Beller (1999). See also Becker (2018).

"realistic," "antirealist," and "antirealistic." These terms have a precise meaning in the philosophy of science, a meaning that seems to be completely unfamiliar to most physicists. And it is not just that these physicists misuse these terms, it is rather that they simply toss them around with no attached meaning at all. This has had terrible consequences for discussions in foundations of quantum theory.

In the proper meaning of the term, *physical theories* are neither realist nor antirealist. That is, as we used to say, a category mistake. It is *a person's attitude toward a physical theory* that is either realist or antirealist. For example, was Copernicus's theory of the structure of the solar system realist or antirealist? That question has no content. The theory was what it was: It postulated that the various planets and the earth engaged in particular sorts of motions. When Osiander wrote the preface to *De Revolutionibus*, he strongly advocated taking an antirealist attitude toward the theory: Don't regard the theory as literally true, but just instrumentally as a convenient way to make certain predictions. He did this to protect Copernicus from the wrath of the Catholic church. Copernicus himself, and Galileo, adopted the opposite attitude: They wanted to argue that the theory is literally true, by reference to its explanatory power and simplicity. And they inherited certain physical problems (for example, problems in terrestrial mechanics) because of their attitude. But the theory toward which Osiander was antirealist and Galileo realist is one and the same theory. The theory itself is neither.

The scientific realist maintains that in at least some cases, we have good evidential reasons to accept theories or theoretical claims as true, or approximately true, or on-the-road-to-truth. The scientific antirealist denies this. These attitudes come in degrees: You can be a mild, medium, or strong scientific realist and similarly a mild, medium, or strong scientific antirealist. Ultimately, this is a question addressed by epistemology and confirmation theory. But this book is not about either epistemology or confirmation theory, so the issue of whether one should be a scientific realist or antirealist, and to what degree, is never even broached. Like "Copenhagen Interpretation," the very terms "realist" and "antirealist" do not appear outside this Introduction.

The real damage that has been done by misapplying the term "realist" to *theories* rather than to *people's attitude toward theories* is raising false hopes. For example, we will see that Bell's theorem, together with reported data, rules out the possibility of any empirically adequate physical theory that is local in a precise sense of the term "local." The Pusey, Barrett, and Rudolph (PBR) theorem, together with data that matches the predictions of quantum theory, rules out the possibility of any empirically adequate "psi-epistemic" physical theory. But often, when reporting these crucial results, the term "realist" or "realistic" is snuck in. Bell, we are told, ruled out all local *realistic* theories, for example. And that locution strongly suggests that one can avoid nonlocality and evade Bell's result by saying that *realism* is what ought to be abandoned. But this suggestion is nonsensical. Bell proves that no local theory, full stop, can predict violations of his inequality. Whether some person's attitude toward the theory is one of scientific realism or not is neither here nor there. If I had my druthers, "realist" and "anti-realist" would be banned from these foundational discussions. And in my own book, I have my druthers, so I will not mention these terms again.

I owe an immense debt of gratitude to many people who have devoted their energy to improving this book. I received tremendously helpful comments from Chris Meacham, Chisti Stoica, Dan Pinkel, Bert Sweet, two anonymous referees, and students in my graduate seminar at New York University who were used to test-drive an earlier version. Zee Perry kindly turned some of my primitive images into polished figures: It will be obvious which is which. Cyd Westmoreland did a splendid job copyediting the manuscript.

I would never have been able to approach this project if not for years of discussion with David Albert, Detlef Dürr, Barry Loewer, the late Robert Weingard, Nino Zanghí, and above all Shelly Goldstein, to whom this feeble attempt is dedicated.

Neither this book, nor anything else of value in my life, would exist if not for Vishnya Maudlin. What she has given is beyond measure and description and can never be adequately acknowledged with mere words.

Philosophy of Physics

CHAPTER 1

Eight Experiments

Physics has traditionally been characterized as the science of matter in motion. Rough as this characterization is, it illuminates the standing of physics with respect to all other empirical sciences. Whatever else the objects of the various empirical sciences are, they are all instances of matter in motion. Every biological system, every economic system, every psychological system, every astronomical system is also matter in motion and so falls under the purview of physics. But not every physical system is biological or economic or psychological or astronomical. This is not to argue that these other empirical sciences reduce to physics, or that the other sciences do not provide an understanding of systems that is distinct from a purely physical account of them. Still, physics aspires to a sort of universality that is unique among empirical sciences and holds, in that sense, a foundational position among them.

The phrase "matter in motion" presents two targets for further analysis: "matter" and "motion." Present physics elucidates the "motion" of an object as its trajectory through space-time. A precise understanding of just what this is requires a precise account of the structure of space-time. The physical account of space-time structure has changed through the ages, and at present the best theory is the General Theory of Relativity. The nature of space-time itself and the geometrical structure of space-time is the topic of the companion volume to this one: *Philosophy of Physics: Space and Time*. The present volume addresses the question: What is matter? The best theory of matter presently available is quantum theory. Our main task is to understand just what quantum theory claims about the nature of the material constituents of the world.

As straightforward as this sounds, we must first confront a great paradox about modern physics. The two pillars on which modern physics rests are the General Theory of Relativity and quantum theory, but the status of these two theoretical systems is completely different. General Relativity is, in its own terms, completely clear and precise. It presents a novel account of space-time structure that takes some application and effort to completely grasp, but what the theory says is unambiguous. The more one works with it, the clearer it becomes, and there are no great debates among General Relativists about how to understand it. (The only bit of unclarity occurs exactly where one has to represent the distribution of matter in the theory, using the stress-energy tensor. Einstein remarked that that part of his theory is "low grade wood," while the part describing the space-time structure itself is "fine marble."[1]) In contrast, no consensus at all exists among physicists about how to understand quantum theory. There just is no precise, exact physical theory called "quantum theory" to be presented in these pages. Instead, there is raging controversy.

How can that be? After all, dozens and dozens of textbooks of quantum theory have been published, and thousands of physics students learn quantum theory every year. Some predictions of quantum theory have been subjected to the most exacting and rigorous tests in human history and have passed them. The whole microelectronics industry depends on quantum-mechanical calculations. How can the manifest and overwhelming empirical success of quantum theory be reconciled with complete uncertainty about what the theory claims about the nature of matter?

What is presented in the average physics textbook, what students learn and researchers use, turns out not to be a precise physical theory at all. It is rather a very effective and accurate *recipe* for making certain sorts of predictions. What physics students learn is how to use the recipe. For all practical purposes, when designing microchips and predicting the outcomes of experiments, this ability suffices. But if a physics student happens to be unsatisfied with just learning these mathematical techniques

[1] Einstein (1950), p. 84.

for making predictions and asks instead what the theory claims about the physical world, she or he is likely to be met with a canonical response: Shut up and calculate!

What about the recipe? Is it, at least, perfectly precise? It is not. John Stewart Bell pressed just this complaint:

> A preliminary account of these notions was entitled 'Quantum field theory without observers, or observables, or measurements, or systems, or apparatus, or wavefunction collapse, or anything like that'. That could suggest to some that the issue in question is a philosophical one. But I insist that my concern is strictly professional. I think that conventional formulations of quantum theory, and of quantum field theory in particular, are unprofessionally vague and ambiguous. Professional theoretical physicists ought to be able to do better.[2]

Bell's complaint is that the predictive recipe found in textbooks uses such terms as "observer" and "measurement" and "apparatus" that are not completely precise and clear. This complaint about quantum theory does not originate with Bell: Einstein famously asked whether a mouse could bring about drastic changes in the universe just by looking at it.[3] Einstein's point was that some formulations of quantum theory seek to associate a particular sudden change in the physical state of the universe ("collapse of the wavefunction") with acts of observation. If this is to count as a precise physical theory, then one needs a precise physical characterization of an observation. As Bell put it: "Was the wavefunction of the world waiting to jump for thousands of millions of years until a single-celled living creature appeared? Or did it have to wait a little longer, for a better qualified system . . . with a Ph.D.?"[4]

These imprecisions in the formulation of the quantum recipe do not have noticeable practical effects when it comes to making predictions. Physicists know well enough when a certain

[2] Bell (2004), p. 173.
[3] Reported by Hugh Everett in Everett (2012), p. 157.
[4] Bell (2004), p. 216.

laboratory operation is to count as an observation, and what it is an observation of. Quantum theory predicts the outcomes of these experiments with stunning accuracy. But if one's main interest is in the nature of the physical world rather than the pragmatics of generating predictions, this ability is of no solace. For the recipe simply does not contain any univocal account of the world itself. To illustrate this, the standard recipe does use a mathematical operation that can be called "collapse of the wavefunction." But if one asks whether that mathematical operation corresponds to a real physical change in the world itself, the recipe does not say. And practicing physicists do not agree on the answer. Some will refuse to hazard an opinion about it.

Bell's complaint might seem incredible. If the problems with quantum theory are not "merely philosophical" but rather consist of the theory being unprofessionally vague and ambiguous as physics, why don't the physics textbooks mention this? Much of the problem has been papered over by a misleading choice of terminology. A standard retort one might hear is this: Quantum mechanics as a physical theory is perfectly precise (after all, it has been used to make tremendously precise predictions!), but the interpretation of the theory is disputable. And, one might also hear, interpretation is a philosophical problem rather than a physical one. Physicists can renounce the desire to have any interpretation at all and just work with the theory. An interpretation, whatever it is, must be just an inessential luxury, like the heated seats in a car: It makes you feel more comfortable but plays no practical role in getting you from here to there.

This way of talking is misleading, because it does not correspond to what should be meant by a physical theory, or at least a fundamental physical theory. A physical theory should contain a physical *ontology*: What the theory postulates to exist as physically real. And it should also contain *dynamics*: laws (either deterministic or probabilistic) describing how these physically real entities behave. In a precise physical theory, both the ontology and the dynamics are represented in sharp mathematical terms. But it is exactly in this sense that the quantum-mechanical prediction-making recipe is not a physical theory. It does not specify what physically exists and how it behaves, but rather gives a (slightly vague) procedure for making statistical predictions about the outcomes of experiments. And what are often called "alternative interpretations of quantum theory" are rather alternative precise physical theories with exactly defined physical ontologies and dynamics that (if true) would explain why the quantum recipe works as well as it does.

Not every physical theory makes any pretense to provide a precisely characterized fundamental ontology. A physical theory may be put forward with the explicit warning that it is merely an approximation, that what it presents without further analysis is, nonetheless, derivative, and emerges from some deeper theory that we do not yet have in hand. In such a case, there may be circumstances in which the lowest level ontology actually mentioned by the theory is not precisely characterized. In the rest of this book, I will treat the theories under discussion as presenting a fundamental ontology that is not taken to be further analyzable, unless I indicate otherwise.

A precisely defined physical theory, in this sense, would never use terms like "observation," "measurement," "system," or "apparatus" in its fundamental postulates. It would instead say precisely *what exists* and *how it behaves*. If this description is correct, then the theory would account for the outcomes of all experiments, since experiments contain existing things that behave somehow. Applying such a physical theory to a laboratory situation would never require one to divide the laboratory up into "system" and "apparatus" or to make a judgment about whether an interaction should count as a measurement. Rather, the theory would postulate a physical description of the laboratory and use the dynamics to predict what the apparatus will (or might) do. Those predictions can then be compared to the data reported.

So far, then, we have distinguished three things: a physical theory, a recipe for making predictions, and the sort of data or phenomena that might be reported by an experimentalist. What is usually called "quantum theory" is a recipe or prescription, using some somewhat vague terms, for making predictions about data. If we are interested in the nature of the physical world, what we want is instead a theory—a precise articulation of what there is and how the physical world behaves, not just in the laboratory but at all places and times. The theory should be able to explain the success of the recipe and thereby also explain the phenomena.

Our order of investigation will start with some phenomena or data. We will try to report these phenomena in a "theory neutral" way, although in the end this will not quite be possible. But, as Aristotle said, any proper scientific investigation should start with what is clearer and more familiar to us and ascend to what is clearer by nature (*Physics* 184a16). We start with what we can see and try to end with an exactly articulated theory of what it really is.

Our phenomena are encapsulated in eight experiments.

Experiment 1: The Cathode Ray Tube

The two ends of an electrical battery are called "electrodes." The positive electrode is the *anode*, and the negative one is the *cathode*. Run wires from these electrodes to two conductive plates, put an open aperture in the anode, place a phosphor-coated screen beyond the anode, and enclose the whole apparatus in an evacuated tube. Finally, add a controllable heating element to the cathode. This apparatus, minus the heating element, was invented by Ferdinand Braun in 1897 and later came to be called a *cathode ray tube* (CRT). The heating element was added in the 1920s by John B. Johnson and Harry Weinhart.

Our first experiment consists of adjusting the heating element so the cathode warms up. When the cathode is quite hot, a bright spot, roughly the shape of the aperture in the anode, appears on the phosphorescent screen (Figure 1a, 1b). As we turn the heating element down, the spot gets dimmer and dimmer. Eventually, the spot no longer shines steadily, but instead individual flashes of light appear in the same area (Figure 1c). As the heat is further lowered, these individual flashes become less and less frequent, eventually only appearing one at a time, with significant gaps between them. But if we keep track of these individual flashes, over time they trace out the same region as the original steady spot.

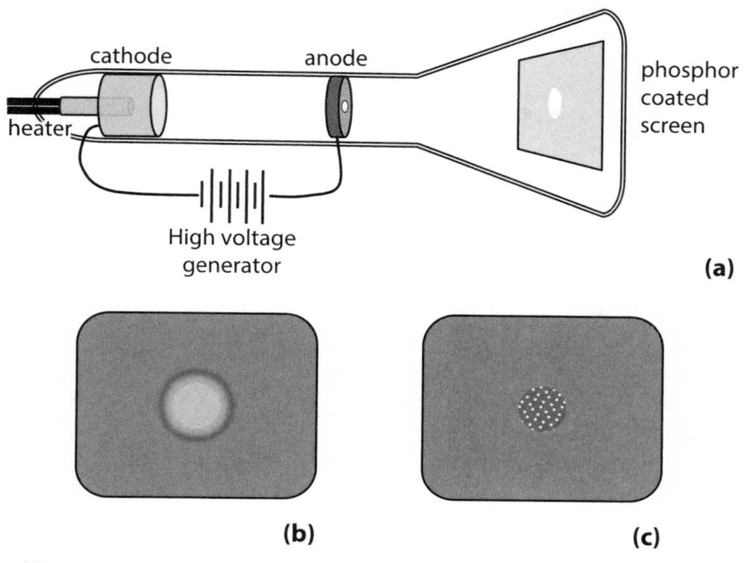

Figure 1

These are the phenomena or data. They immediately suggest certain hypotheses about what is going on inside the tube, but for the moment, we want to distinguish any such hypotheses from the data themselves. The phenomena suggest, for example, that something is going from the cathode (where the heating is applied), through the aperture in the anode, and to the phosphorescent screen. We can test this hypothesis by moving screen toward the anode while the spot is steady. The spot remains steady, and it narrows and brightens as it approaches the anode. Just in front of the anode, the spot is the same shape as the aperture. One can place a screen between the cathode and anode, where it will light with a larger, brighter, more diffuse glow. So there does seem to be something emitted from the cathode and going to the screen. Originally, this something was called *cathode rays*.

When we turn down the heat, these cathode rays exhibit a sort of discrete or grainy character, producing one flash at a time. We could not have predicted this behavior: The spot might have just dimmed uniformly without ever resolving into individual scintillations. These individual discrete events suggest a further hypothesis, namely, that the cathode rays are composed of a collection of individual particles. These hypothetical particles were eventually called *electrons*, and the whole cathode/anode apparatus is sometimes referred to as an *electron gun*.

The model suggested by the term "electron gun" is strengthened by the following fact. Increasing the voltage of the battery increases the "speed" of the cathode rays in the sense that if we measure how long it takes between connecting the battery and seeing the spot, it takes less time for higher voltages. This relation yields a narrative: Heating the cathode boils off electrons, which, being negatively charged, are repelled by the negatively charged cathode and attracted to the positively charged anode. The greater the voltage difference between the two, the faster the electrons will go, with some passing through the aperture in the anode and continuing on to the screen.

It is indeed difficult to resist this particle hypothesis, but for the moment, resist it we must. The postulation of individual particles that travel from the cathode to the screen is not itself part of the data, although it might be part of a theory meant to account for the data.

Skepticism about the physical existence of individual discrete particles in this experimental situation may seem excessively cautious, but our next two experiments point in another direction.

Experiment 2: The Single Slit

If individual particles are flying from the cathode to the screen, then an object placed between the cathode and the screen might be able to affect these particles. As our first test of this hypothesis, we place a barrier with a single slit. The spot on the phosphorescent screen becomes long and thin, much as one might have anticipated (Figure 2a). Making the slit thinner in what we will call the z-direction initially makes the image thinner, again as one would expect. But beyond a certain point, a peculiar

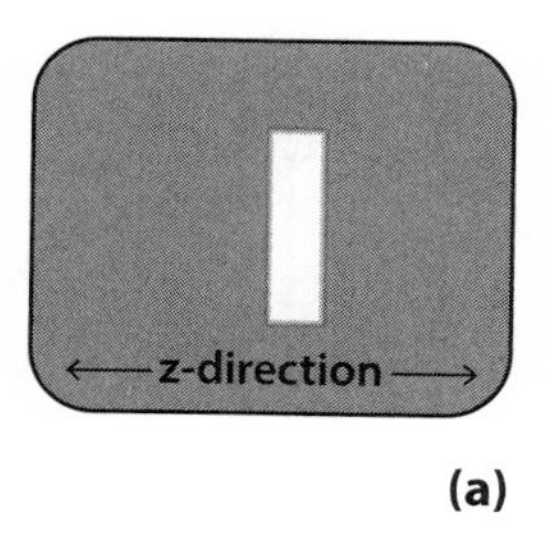

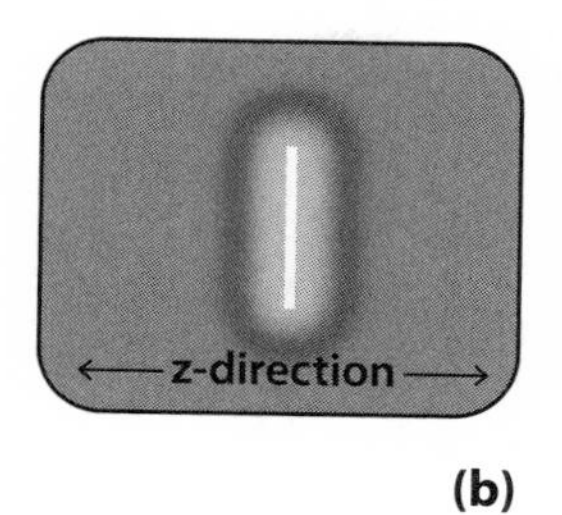

Figure 2

thing happens: making the slit even thinner results in the spot becoming wider and more spread out in the *z*-direction (Figure 2b). (In addition, the image starts to show some variation of brightness, with dark patches emerging. We leave that aside for now).

Our initial hypothesis of particles would not have hinted at this new development, but it is reminiscent of the familiar behavior of waves called *diffraction*. When a series of plane water waves hit a wide gap in a barrier (wide relative to the wavelength, i.e., distance between the crests), the wave train that gets through continues largely in the same direction, with just a little dissipation around the edges. But when it hits a very narrow gap, it creates a circular wave pattern on the other side that spreads farther upward and downward (Figures 3a and 3b). Crests are indicated by solid lines and troughs by dotted lines.

Since diffraction occurs when the size of the hole is small compared to the wavelength of the wave, the diffraction can be reduced by shortening the wavelength. And we find that the diffraction of our cathode rays is reduced as we increase the voltage between the cathode and the anode. So in this respect, our cathode rays behave somewhat like water waves, with the wavelength going down as the voltage goes up.

But it is still also the case that as we turn the heating element down, the glow goes from a steady state to a series of individual flashes, so in this sense, the phenomena suggest individual particles. The fact that the cathode rays (or electrons) produce phenomena associated with waves and also phenomena associated

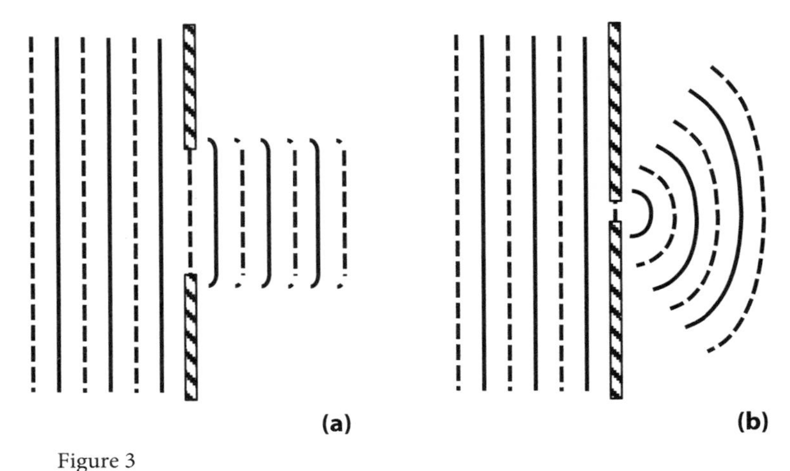

Figure 3

with particles is called *wave-particle duality*. But that is just a description of the phenomena, not an explanation of them.

Experiment 3: The Double Slit

We are now in a position to describe the experiment most often associated with quantum theory: the double-slit experiment. In his classic *Lectures on Physics*, Richard Feynman is referring to the two-slit experiment when he says:

> We choose to examine a phenomenon which is impossible, *absolutely* impossible, to explain in any classical way, and which has in it the heart of quantum mechanics. In reality, it contains the *only* mystery. We cannot explain the mystery in the sense of "explaining" how it works. We will *tell* you how it works. In telling you how it works we will have told you about the basic peculiarities of all quantum mechanics.[5]

[5] Feynman, Leighton, and Sands (1975), Section 37-1.

Feynman is not correct when he says that there is no way to explain this phenomenon in a "classical" way, at least in one sense of "classical." But this much is certainly true: the phenomenon is quite unexpected, and one does not really understand any physical theory that purports to account for the behavior of matter unless one understands how the theory accounts for this phenomenon. Belying Feynman's pessimism, we will discuss several quite different exact physical theories, all of which can explain it.

Experiment 2 already demonstrates a behavior of cathode rays similar to that of water waves: diffraction. But an even more striking characteristic is associated with waves, namely, *interference.* Waves interfere because when they meet each other, they interact by *superposition.* For example, if the crest of one wave arrives at the same place as the equally deep trough of another, they cancel each other out; and if a crest meets a crest, they add to make a crest twice as tall. In Figures 3a,b, the solid lines represent the crests of the water waves, and the dotted lines represent the troughs. Now suppose instead of one hole or slit in the barrier we put two, and suppose that each slit is narrow enough to cause a lot of diffraction: in essence, each slit becomes the source of a set of circular wave patterns emanating from it (Figure 4). Where the crests of one wave meet the troughs of the other, they cancel out, and the water becomes still; where two crests meet or two troughs meet, the water is extremely agitated. Regions where crests coincide with crests and troughs with troughs are indicated by unbroken arrows, and regions where crests meet troughs by broken arrows. This superposition results in *interference bands* at the screen: regions of extremely high activity alternating with quiescent regions. Points on the screen where the difference in distances to the two slits is half a wavelength, or one and a half, or two and a half, and so forth (so the two arriving waves are out of phase) show little wave activity, and points where the difference is an integer number of wavelengths show lots of activity. Following the analogy with diffraction, then, we would expect alternating light and dark bands on the screen. This is indeed exactly what happens. Using a photographic emulsion and turning the heating element down yields a situation in which only individual dots

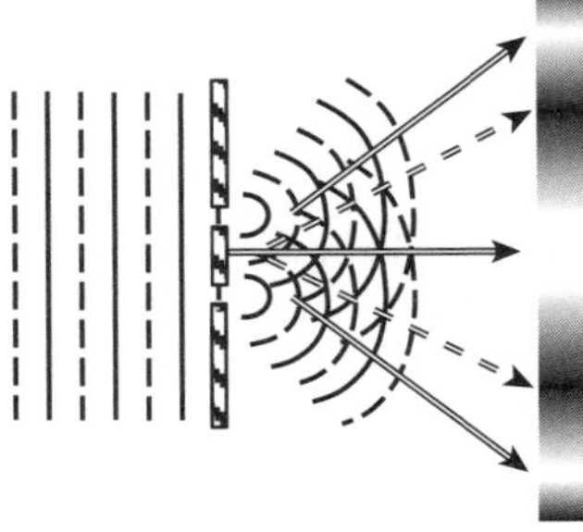

Figure 4

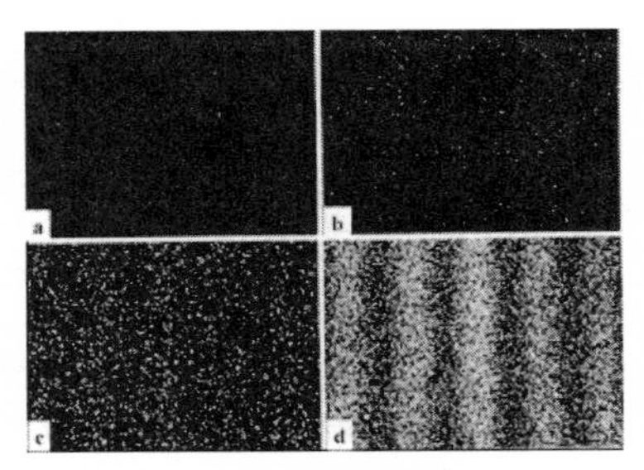

Figure 5. Credit: Reprinted courtesy of the Central Research Laboratory, Hitachi, Ltd., Japan.

appear on the screen, slowly accumulating to form the interference bands (Figure 5).

The two-slit interference experiment simultaneously displays properties we would naturally associate with particles (the individual discrete flashes or dots) and also properties we would naturally associate with waves (the interference bands). This is startling. But why would Feynman make the much stronger statement that the phenomenon cannot possibly be explained in any classical way?

Feynman's idea seems to be this: "to explain in a classical way" means to postulate the existence of individual particles that make their way, along one continuous path, from the cathode to the screen. Each such particle would therefore have to either pass through one slit or pass through the other (or loop around somehow to pass through both). Feynman calls the claim that each particle passes either through one slit or through the other "Proposition A." He then argues that Proposition A has some empirically testable consequences that turn out to be false, showing that we cannot accept it.

Suppose that each cathode ray that reaches the screen passes either just through the upper slit or just through the lower (leaving aside more rococo possibilities). Feynman reasons as follows. We can determine the final distribution of rays that pass through the upper slit by closing off the lower slit and seeing what happens. But we already know what happens: that is just Experiment 2. Similarly, we can close off the upper slit, in which case we get the same spread-out pattern just shifted over a bit. But the gaudily streaked interference pattern of the two-slit experiment is not the sum of these two experiments. Indeed, there are particular locations on the screen where spots will form if only the upper slit is open and spots will form if only the lower slit is open, *but no spots will form if both slits are open.*

Does it really follow from this observation, as Feynman suggests, that Proposition A cannot be true? In chapter 5, I will present a precise physical theory according to which each particle goes through exactly one slit and the interference bands only form when both slits are open. So there are ways to account for the data while validating Proposition A. What Feynman really seems to have in mind is not merely Proposition A, but also the additional proposition (call it Proposition B) that if an electron goes through one slit, then its later behavior will be the same regardless of whether the other slit is open. It is only with this second principle in place that one could infer that, given Proposition A, the distribution of flashes with both slits open would be the sum of the distributions with only one slit open. But Proposition B is not a proposition of classical physics, classical probability theory or classical logic. And the simple fact that there are locations on the screen where flashes occur if only one slit (whichever one) is open but never occur with both open already proves that *for each individual flash, the physical situation at the screen is sensitive to the condition of both slits.* This cannot be denied. What we want is a clear physical account of how it happens.

Denying Propositions A or B suggests that, in some sense, each electron or cathode ray interacts with both slits. And if this is true, then it is not surprising that the behavior at the screen can be sensitive to the fact that both slits are open. But for the electron or cathode ray to interact with both slits, it must somehow be spread out over a region large enough to encompass both slits, just as a water wave would have to be spread out that much. And in that case, the mystery is not so much how the behavior can depend on the state of both slits, but rather why the flashes on the screen occur at discrete, definite localizations. If a cathode ray really spreads out as a water wave does, how do the individual flashes manage to form?

As we will see, different precise theories embrace different horns of this dilemma. In some, the electron unproblematically goes through exactly one slit on its way to the screen, and the trick is to see how the other slit being open affects its later behavior. In others, the electron in some sense goes through both slits, and the trick is to account for the discrete flashes. But the situation is even more complicated: A slight modification of this experiment holds more surprises.

Experiment 4: The Double Slit with Monitoring

Since our main puzzle concerns which slit, if either, the electron goes through on its trip from the anode to the screen, one might well ask: Why not just check? "Checking" means adding some new element to the experimental set-up designed to yield which way information about the electron, that is, information about which slit the particle went through. We will now explore a somewhat unrealistic and idealized modification of the experiment, but the effect of the modification on the phenomena is firmly based on quantum-mechanical principles.

With the thought that the electron is negatively charged, and that negatively charged particles attract positively charged ones, we might hit on the following scheme. Make a small, thin chamber in the screen between the two slits, and place a proton in a position exactly between the slits. Line the ends of the chamber with a substance that will emit a flash if a proton is absorbed (Figure 6).

If the electron goes through the upper slit, the proton will be attracted upward and the flash will occur at the top of the chamber, and if the electron goes through the lower slit, the flash will occur at the bottom. We can check the reliability of this monitoring mechanism by running it first with each slit closed. If it is 100% reliable, there will be a flash in the corresponding part of

Figure 6

the chamber when and only when there is a flash on the screen. That is, with the lower slit closed, there is a flash in the upper part of the chamber exactly when there is a flash on the screen, and similarly mutatis mutandis with the upper slit closed. We could also imagine less than perfect reliability: The corresponding flash might only happen 75% of the time, for example. We will consider this possibility presently.

Supposing we have achieved 100% reliability with only one slit open, what will happen when both slits are unblocked? Naively, we might expect to see the interference bands, as in Experiment 3, but now with additional information from the flash in the chamber about which slit the electron went through. Or, if the electron somehow goes through both slits and so would equally pull the proton up and down, maybe the proton will just remain symmetrically in the middle. Experiment 3 gives us no clue about the outcome.

As it turns out, this is what happens: Our completely reliable monitoring mechanism will continue to signal one slit or the other when there is a flash on the screen. And over time, about half of the electrons will be "seen" to have gone through the upper slit and about half through the lower. *But the interference bands will completely disappear*. The distribution of flashes on the screen will now be a simple sum of the distributions that occur when only one of the two slits is open. To put it somewhat poetically, when the path of the electron through the apparatus is observed, the behavior of the electrons changes from being wavelike (showing interference) to being particle-like (showing no interference). But notice that the "observer" in this poetic description is not even as sophisticated as a mouse. It is just a single proton whose own behavior has been coupled in the right way to that of the electron. There is something about that physical coupling that both destroys the interference and also seems to yield information about what the electron did.

What if we weaken the coupling between the electron and the proton? Suppose, for example, instead of reacting perfectly reliably when an electron goes by, the proton only moves from the central position 75% of the time (but always in the right direction, as checked with only one slit open)? What will we see then?

As the reliability of the monitor is reduced, the interference bands will slowly and continuously emerge. But as long as the behavior of the proton is correlated with the electron (with only one slit open) the interference bands will not be as strong as in Experiment 3. And the role played by the proton in destroying the interference is illustrated in a very striking way. If one divides the electron flashes on the screen into those that occur when the proton gives a result and those that occur when the proton stays in the central part of the chamber, the washed-out interference bands get split into two strikingly different sets. In the set where the proton indicates a slit, there is no interference at all, and in the set where it remains in the center, there is full interference. The total distribution is just the sum of these. As we progressively weaken the coupling with the proton, the interference bands progressively reemerge to full force.

One might well wonder how any clear and precise physical account of what is going on could yield this sort of behavior. What sort of pattern appears on the screen seems to depend on whether, in some sense, anyone or anything is "watching" the electron. But must the physical theory therefore define "watching" in order to be articulated? How can that be done? Has the observer somehow claimed a central place in physics? Many physicists over the years have drawn just this conclusion. Experiment 4 gives us some indication of the phenomena that led them to it. But the very simplicity of the "watcher" in this experiment is promising. There is little prospect of producing an exact physical characterization of something as large and complicated as a mouse. But a single proton, coupled by electrical attraction to an electron, is exactly the sort of thing we expect an exact physical theory to treat with complete precision. So Experiment 4 ought to give us some hope.

Experiment 5: Spin

Our previous experiments have illustrated some of the peculiarities of quantum theory. It is easy to see in these phenomena wave-particle duality, since individual flashes are suggestive of particles, and the collective interference patterns are suggestive of waves. Our last experiment illustrates how the physical role of observation might appear as a central theme. Even the simple single-slit diffraction experiment provides an instance of the famous Heisenberg uncertainty relations. Werner Heisenberg noticed that as our predictive abilities become better in some ways, they simultaneously become worse in others. The sorts of predictions that trade off in this way are called "complementary." One standard example of complementarity is position and momentum in a given direction. Narrowing the slit in Experiment 2 decreases uncertainty about where in the z-direction a particle that passes the slit will show up just beyond the slit, but it simultaneously increases uncertainty about its z-momentum (i.e., how fast and in what direction it is moving in the z-direction) at that point. This increased spread in possible z-momenta results in the widening of the image in the z-direction far from the slit. But so far we have not had much indication of the "quantum" in quantum theory. It is popularly thought that in quantum theory everything is quantized into discrete units. But in our examples so far, that is not so. Our cathode rays can appear as flashes at any location on the screen, for example.

The simplest physical property that exhibits quantization is called "spin," and manifests itself as an intrinsic angular momentum of a particle. In classical physics, a spinning charged particle has a magnetic polarization. If an object has a north and south magnetic pole, then it will be deflected when travelling through an inhomogeneous magnetic field. Figure 7 shows a diagram of a Stern-Gerlach apparatus that produces this effect.

The apparatus is just a magnet, but because of the asymmetric geometry, the north pole creates a locally stronger magnetic field than does the south pole. A bar magnet in the field oriented with its north pole up and its south pole down would be pushed down, since the north will be repelled by the upper field more strongly than the south is repelled by the lower. Similarly, a bar magnet oriented the opposite way will move up, since the attraction of its south pole upward will overbalance the attraction of the north pole downward. A horizontally oriented bar magnet will not be pushed or pulled either way.

Our electrons are negatively charged, but it does not immediately follow that they have magnetic moments. Classically, a spinning electric charge does create a magnetic field. Hence an intrinsic magnetic moment of a particle is associated with its "spin," irrespective of whether it originates in the actual spinning motion of anything. One way to check for such a magnetic moment is to pass a particle through a Stern-Gerlach apparatus to see whether it is deflected.

If one does this sort of experiment on our cathode rays, the outcome is somewhat unexpected.[6] Every electron is deflected either up or down, with none going straight through. Furthermore,

[6]The physics here is somewhat idealized, although again the basic principles are correct. In practice, this experiment was first done with silver atoms.

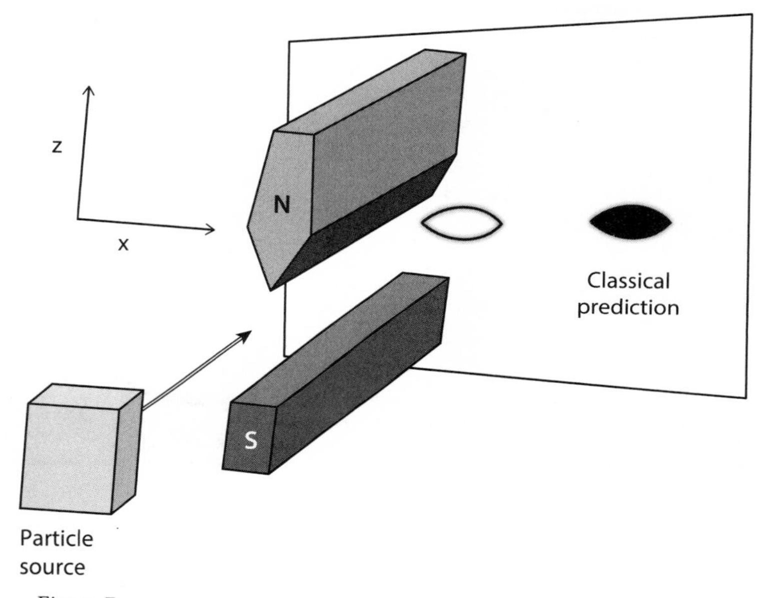

Figure 7

the amount of deflection is exactly the same in both cases. Our electron beam divides into two equally strong parts, one deflected toward the north pole of the apparatus (up-spin) and the other toward the south pole (down-spin). Or, more precisely, our steady lighted patch on the screen splits into two equally bright patches, one above and the other below the midline. And as we turn the beam intensity down, we again get individual flashes, about half in a small region in the upper part of the screen, half in an equally small region in the lower part. Particles in a beam that splits into exactly two parts are called "spin-1/2" particles. If it were to split into three parts, one going straight through, the particles would be spin-1 particles. A beam of spin-3/2 particles splits into four parts, and so on.

In Figure 7, the image on the screen looks like an eye, because the electron beam out at the edges does not go through an inhomogeneous field and so travels straight through. Stern and Gerlach's

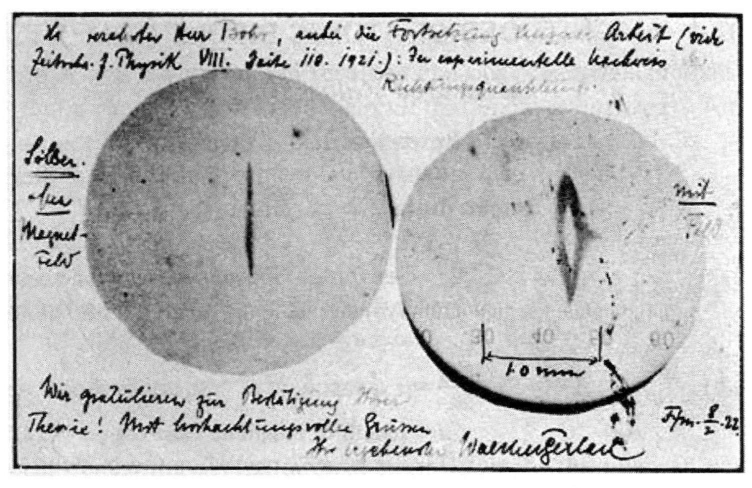

Figure 8. With permission of Niels Bohr Archive, Copenhagen.

actual data are shown in Figure 8 on a postcard that Gerlach sent to Niels Bohr. In our schematic diagrams, we will cut off the sides, so only the most separated parts of the beam are indicated.

The Stern-Gerlach apparatus is itself oriented in some spatial direction. Figure 7 designates the vertical direction as the z-direction and the horizontal one as the x-direction. If we twist the apparatus from the z-orientation to the x-orientation, the beam comes to split in the x-direction, as in the postcard. The apparatus can be set to have any spatial direction.

Since about half of the particles are deflected up and half deflected down, one naturally wonders whether some feature of each individual particle determines which way it goes. It not obvious how to resolve this question experimentally, but some additional experimental configurations are clearly relevant. Let a first Stern-Gerlach apparatus be oriented in the z-direction, splitting the beam into an upper and lower branch. Then place a second apparatus, also oriented in the z-direction, in each of these beams (Figure 9a). We might expect each beam to split again, but it does not: the whole upper beam is deflected up and the whole lower

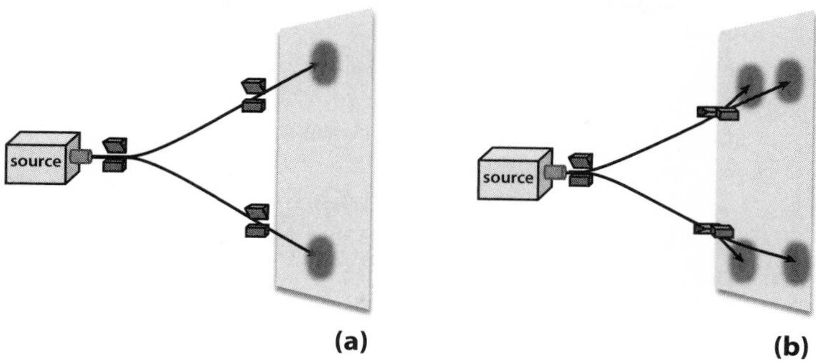

Figure 9

beam deflected down. So electron beams can be prepared so that each electron in them is disposed to be deflected in a particular way by a z-oriented apparatus. Those disposed to be deflected upward are called "z-spin up" electrons, and those disposed to be deflected down called "z-spin down."

We can also follow a z-oriented apparatus with x-oriented ones (Figure 9b). In this case, each beam splits 50-50. So preparing a beam so it can be predicted which way each electron will be deflected in the z-direction results in complete uncertainty about how it will be deflected in the x-direction. And testing the output of the x-apparatus with yet another z-apparatus reveals that the original preparation has been lost: the beam once again splits 50-50 up and down.

All electrons in a z-spin up beam get deflected up in the z-direction and only half do in the x-direction. What if we slowly rotate the second apparatus from the z-orientation to the x-orientation? Unsurprisingly, the proportion deflected in the up direction (with respect to the apparatus) varies smoothly from 1 to 0.5. More quantitatively, the proportion deflected up at the second apparatus is $\cos^2(\theta/2)$, where θ is the angle between the orientations of the two apparatuses.

Our spin experiments illustrate the quantization of spin, since each electron responds to the experimental condition in one of two possible ways. They also illustrate the Heisenberg uncertainty relations: The more certain it is how an electron will react to a z-oriented apparatus, the less certain it will be how the electron reacts to one that is x-oriented and vice versa. This is analogous to the situation with z-position and z-momentum in the single-slit experiment.

The quantization of spin offers particularly sharp and clear experimental possibilities that are the subject of our next laboratory configuration.

Experiment 6: The Interferometer

Our next experiment refines some of the phenomena we have already discussed. We have seen how a Stern-Gerlach apparatus can split an incoming beam of spin-1/2 particles into two beams. Those beams can be further manipulated and recombined in an experimental configuration that was originally developed for light by Ludwig Mach and Ludwig Zehnder, and hence is known as the Mach-Zehnder interferometer.

The first experiment is a slight variation on a spin experiment we have already discussed. Prepare an x-spin up beam of electrons and pass it through a z-oriented Stern-Gerlach device. We have already remarked that if we pass either of the output beams through an x-oriented apparatus, the beam will again split: apparently the z-oriented magnet "scrambles" the information about the prepared x-spin. In itself, this is not terribly surprising. The interaction of the beam with the new magnetic field could have all sorts of disruptive effects. But the Mach-Zehnder configuration allows us to steer the output beams of the z-oriented device back together, having been widely separated from each other for some time (Figure 10). A natural train of thought runs as follows: The x-spin of each separate output beam of the z-oriented magnet has been scrambled, with each particle equally likely to be deflected up or down. When two such scrambled beams are combined, the result should be just as scrambled. So the recombined beam should also be equally split if passed through an x-oriented magnet at point A in the figure.

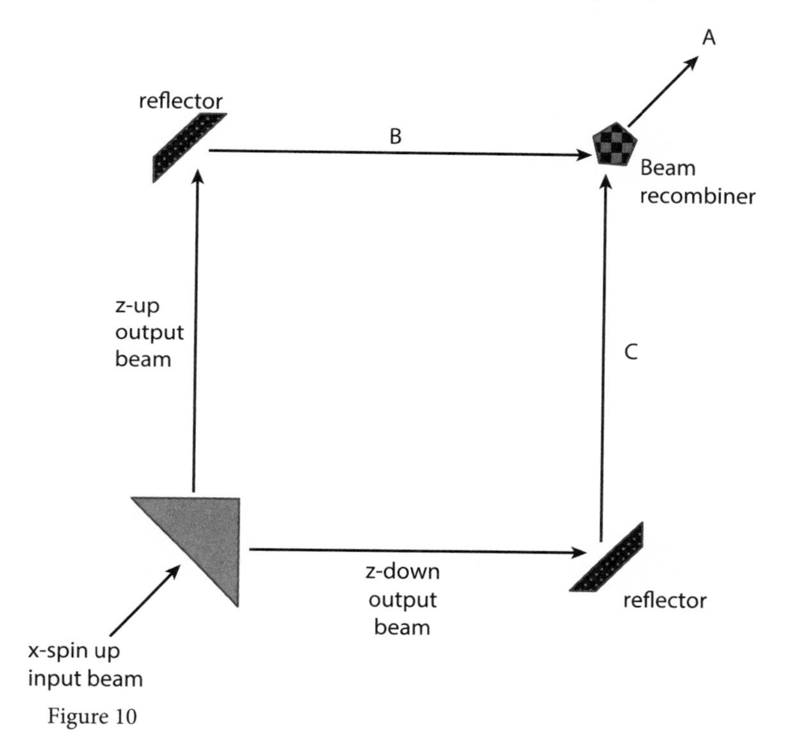

Figure 10

This, however, is not what happens. Every single electron is deflected up at point A even though half would have been deflected down if the beam had instead been checked by a pair of devices located at points B and C. And if x-spin down electrons are fed into the interferometer, we get a similar result: Half will be deflected down if the x-spin is checked at points B and C, but all are deflected down if the x-spin of the recombined beam is checked at A. Information about how the original beam was prepared is somehow transmitted through the splitting and recombination, even though that very information appears to have been lost half way through! And once again, these statistics hold even if we turn down the intensity of the incoming beam so only one electron goes through the interferometer at a time.

The interferometer set-up allows us a new opportunity: to intervene on one branch of the split beam before it is recombined. One intervention is particularly striking. We have already seen that the spin of an electron is like an intrinsic magnetic moment, akin to that of a classical spinning charge. In a classical theory, applying a magnetic field to such an object will cause it to precess (i.e., to slowly rotate in space). One sort of magnetic field, applied for the right amount of time, would cause the object to rotate though a full 360° and hence (apparently) return to the same state it started in. We can apply such a magnetic field to our electrons,[7] and check that the precession does return the beam to its initial state: an x-spin up beam remains x-spin up when the magnetic field is applied, an x-spin down beam remains x-spin down, z-spin up remains z-spin up, etc. Further, an x-spin up beam gets converted to x-down if the magnetic field is applied for half the time, just as one would expect if it were rotated through 180°. A device that applies the magnetic field for the full time is an example of what David Albert calls a "total of nothing box" because the observable statistics of any beam are unchanged by the application of the magnetic field.[8] As far as predictions are concerned, a beam that has had the magnetic field applied behaves just like one that has not. The foregoing remarks hold so long as the whole beam is subjected to the magnetic field.

But suppose we split the beam in the interferometer and apply the magnetic field to only one part of the split beam (at point B, for example) and then recombine the beams. This intervention has a dramatic effect on the outcome. Without the magnetic field, as we have seen, if we feed an x-spin up beam in, we get an x-spin up beam out after the recombination. But with the magnetic field in place, when we feed an x-spin up beam in, we get an x-spin *down* beam out. Every single electron is deflected down by an x-oriented magnet at the end, while without the magnetic field, every single electron is deflected up. In other words, every

[7] I am fudging the actual physics a bit: The experiment described here was carried out on neutrons rather than electrons. Neutrons also are spin-1/2 particles.

[8] Albert (1992), p. 11.

electron fed through our device is demonstrably sensitive to the physical conditions along both paths in the interferometer: A certain magnetic field applied either at point B or at point C (but not both) will alter the behavior of every electron that passes through.

This is not, strictly speaking, a new sort of observation. We have already seen in the two-slit interference that every electron is sensitive to the state of both slits. The Mach-Zehnder configuration brings out this fact in a particularly striking way, since the two paths through the interferometer can be made to diverge from each other by arbitrary distances. Nonetheless, an intervention on either branch can have an effect on every single electron.

Experiment 7: The EPR Experiment

Unlike the interferometer, our final two experiments bring in fundamentally new features of quantum theory. Indeed, we are starting on the path to the most puzzling and astonishing physical phenomena predicted by the quantum formalism and verified in the laboratory. These phenomena essentially involve collections of particles rather than single particles or beams of single particles. So far, only Experiment 4, the Double Slit with Monitoring, has required more than one particle at a time. Experiment 4 demands this because the monitoring proton and passing electron must interact for the monitoring to occur. We now embark on a deeper investigation into such interactions, and into the information that the behavior of one particle can yield about another.

The first experiment is a modification, proposed by David Bohm, of an experimental situation described by Albert Einstein, Boris Podolsky, and Nathan Rosen in "Can Quantum-Mechanical Description of Reality Be Considered Complete?" (1935), now known as the EPR paper.[9] In that paper, the discussion concerned the positions and momenta of a pair of particles prepared in a special state. Bohm changed the example to use spin in

[9]Einstein, Podolsky, and Rosen (1935).

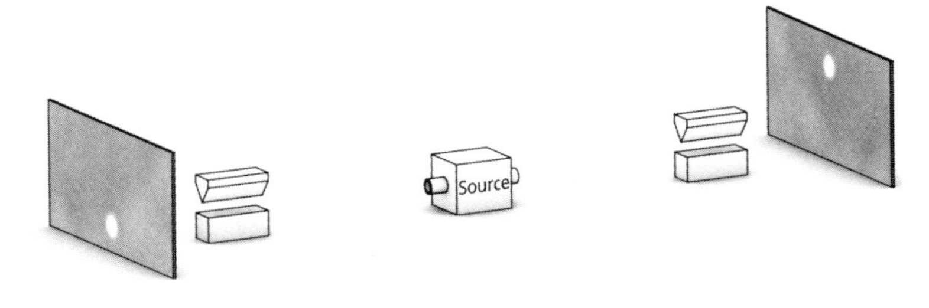

Figure 11

various directions rather than position and momentum, and we will follow his simpler example.

Unlike all the experiments described so far, the basic phenomena in the EPR experiment seem unremarkable. A pair of electrons is prepared in a particular quantum-mechanical state (called the "singlet" state) and is allowed to separate to an arbitrary distance from each other. Each electron is then passed through a Stern-Gerlach apparatus oriented in a specific spatial direction (Figure 11). For example, both of the devices might be oriented in the *z*-direction. In this case, the two electrons always behave in opposite ways: If one is deflected upward in the apparatus, the other is deflected downward. Therefore, by observing how one electron behaves, one can predict with perfect accuracy how the other will (or has).

So what is so remarkable about this? It is true that the behavior of one electron provides information about how the other will behave, but everyday instances of these sorts of correlations are commonplace. John Bell used the amusing example of the physicist Reinhold Bertlmann, who always wore socks of different colors.[10] Given this somewhat idiosyncratic choice of how to get dressed, the color of one sock (pink, say) provides information about the color of the other (not-pink).

[10]"Bertlmann's Socks and the Nature of Reality," reprinted as Chapter 16 in Bell (2004).

But notably, this prosaic account of the phenomenon essentially presupposes that the socks have their colors all along, from the time Bertlmann got dressed. If, somehow, neither sock had any definite color in the morning, if a sock only acquired a definite color some time later (when observed, for example), then the situation would be truly remarkable. It would be remarkable first because of the no-definite-color to definite-color transition. One would rightly wonder about the physics of that change. But even granting that, there is a residual surprise, for not only would the one sock have to somehow come to become actually pink at some point, but the other sock (which might be miles away) would also somehow have to become some color other than pink, so that the colors would always be different. This idea, that interacting with one sock can somehow not merely provide information about the other but actually affect the physical state of the other, is an example of the possibility of *quantum nonlocality*.

Einstein, Podolsky, and Rosen never took the possibility of such a nonlocal physical interaction between the socks (or the electrons) seriously. In fact, they thought the idea so absurd that they never imagined anyone would entertain it. What the EPR article pointed out was that to avoid such a strange "spooky action-at-a-distance" (in Einstein's famous phrase), one has to postulate that the two electrons described above have definite dispositions concerning how they would react to the magnets from the moment they are produced and separate from each other. One of the electrons has to be z-spin up and the other z-spin down from the outset. Otherwise, how could either be sensitive to the behavior of the other in the right way to preserve the perfect anticorrelation?

It is worthwhile to belabor this point a little. Imagine, as an analogy, that you and a friend are going to be subjected to the following ordeal. You are going to be taken into separate rooms and asked a yes-no question. If you give different answers to the question, you will both be let go, but if you give the same response you will both be punished. You have absolutely no idea what the question will be.

You would likely not be daunted by this ordeal. After all, there is a simple way to avoid the punishment. You just have to agree to give different answers to the question. But to carry out this scheme, you must do more than just agree to give different answers, you must agree, while you can still communicate, exactly which answer each one will give. If your friend merely suggests that you give different answers but then leaves before saying how he will answer, then you are no better off than before. Without knowing how he will answer, you have no means to arrange your answer to be different, no matter how much you want it to be. Unless you somehow later acquire information about how your friend has answered, your strategy will be useless.

Similarly, if neither electron has a definite disposition to be deflected either up or down by the Stern-Gerlach apparatus when they separate, then it is hard to see how they can be assured of being deflected in opposite ways without some physical mechanism that makes one sensitive to what the other does. And since the electrons can be carried arbitrarily far apart, such a mechanism would have to work at arbitrary distances. Einstein never accepted the physical reality of such a mechanism, and he concluded that the electrons had to have their dispositions all along.

This conclusion in itself might seem rather obvious and mild. But everything we have said about pairs of electrons in the singlet state and z-oriented Stern-Gerlach magnets holds as well for the electrons and x-oriented magnets, or y-oriented magnets, or magnets oriented in any spatial direction. So if we conclude that, to avoid the spooky action-at-a-distance, each electron must have a definite disposition about how it will behave if confronted with a z-oriented magnet, then it must equally have a definite disposition with respect to x-oriented magnets, y-oriented magnets, and so on. But we have already seen that we can't prepare a beam of electrons so that we can both predict with certainty how each electron will react to a z-oriented magnet and how it will react to an x-oriented magnet—that impossibility is an example of the Heisenberg uncertainty relation. Nonetheless, if we are to avoid Einstein's spooky action-at-a-distance, each electron in a singlet state must have a definite propensity to react a particular way to a z-oriented magnet and to an x-oriented magnet.

There is no contradiction in saying that on one hand, it is impossible to prepare a beam of electrons so that all will be deflected up if confronted with a z-oriented magnet and all will be deflected down if confronted with an x-oriented magnet, while on the other hand insisting that individual electrons have both these propensities. But if such individual electrons exist, standard quantum theory does not have the resources to represent their physical state. That was the issue as the EPR paper presented it: Is the quantum-mechanical description of a system complete (i.e., does it somehow represent all physical characteristics of the system)? Having rejected the spooky action-at-a-distance, EPR conclude that some physical characteristic of each electron must determine how it would behave in all these different experimental conditions, and therefore the quantum-mechanical description of the individual system is not complete. But as far as logic goes, one could reject their conclusion by embracing the notion of action-at-a-distance.

Einstein did not imagine that his rejection of action-at-a-distance could be subject to experimental test. In 1964, John Bell proved him wrong.

Experiment 8: GHZ/Tests of Bell's Inequality

We have arrived at the strangest and most counterintuitive phenomena predicted by quantum theory and confirmed in the lab. We will mention two related examples of the general phenomenon, one conceptually simpler but experimentally harder, the other experimentally easier to confirm but slightly more complicated to analyze.

The conceptually simpler example was discovered in 1989 by Daniel Greenberger, Michael Horne and Anton Zeilinger, inspired by reflection on Bell's work. The experimental situation they envisage bears obvious similarities to Bohm's spin version of the EPR example, except three particles are involved rather than two. This triple of particles is created in a particular quantum-mechanical state and allowed to separate to arbitrary distances

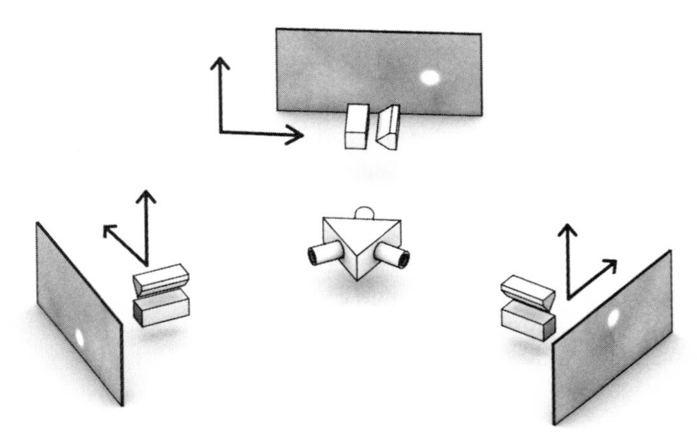

Figure 12

apart, where each will either be subjected to an x-oriented magnet or to a z-oriented magnet (Figure 12). We can imagine the choice between the two experimental arrangements for each particle being made at random in whatever way one likes: by whim, by a random number generator, by flipping a coin, and so forth. The predictions are independent of how this choice happens to be made. Figure 12 depicts two z-oriented and one x-oriented magnets.

If we denote the particles by 1, 2, and 3, then the possible local experimental conditions can be labeled X_1, Z_1, X_2, Z_2, X_3 and Z_3. The global experimental situation in a particular run of the experiment will specify the situation for each of the three magnets, so there will be eight possible global experimental configurations: $X_1X_2X_3$, $X_1X_2Z_3$, $X_1Z_2X_3$, $X_1Z_2Z_3$, $Z_1X_2X_3$, $Z_1X_2Z_3$, $Z_1Z_2X_3$, and $Z_1Z_2Z_3$. If we decide which way to set each apparatus by the flip of a fair coin, then we would expect each of these global conditions to obtain about once in every eight runs of the experiment.

Of these eight possible global configurations, currently we are only interested in four: $X_1X_2X_3$, $X_1Z_2Z_3$, $Z_1X_2Z_3$, and $Z_1Z_2X_3$. After many runs of the experiment, we would notice the following unbroken regularities. When the $X_1X_2X_3$ configuration obtains (i.e., when all three magnets are oriented in the *x*-direction), then an odd number of the particles (either one or all three) are deflected in the "up" direction. But when any of the other three configurations is chosen, so one magnet is in the *x*-direction and the other two in the *z*-direction, an even number of particles (either zero or two) are deflected in the up direction. That is the observed phenomenon.

Why is this so puzzling? First of all, note that in any of the four arrangements, knowing two of the outcomes allows us to predict the third with certainty. In this regard, the situation is similar to the EPR set-up, where the result on one side provides perfect information about the result on the other. And by exactly the same reasoning as in EPR, we conclude that either each particle is always physically disposed to react in a particular way to each possible setting of the magnet, or some spooky action-at-a-distance takes place. For if any one particle did not have a definite disposition about how it would behave, how could the other pair of particles arrange their reactions to maintain the proper statistics unless they were somehow influenced by the unpredictable behavior of the third particle? For example, when particle 1 encounters an *x*-oriented magnet, if it could either be deflected up or be deflected down, with nothing in its physical state determining which, how could the other particles be certain to show the correct number of "up" outcomes if what they do is completely unaffected by what particle 1 does?

So just as in the EPR argument, we have a choice: either the outcomes of the experiments are predetermined locally by the physical states of the individual particles, or there must be some spooky action-at-a-distance. But unlike the EPR case, it is easy to prove that the outcomes can't be predetermined just by the physical state of the particle that reaches each apparatus, independently of what happens to the others. The mathematical proof of this is simple and is wonderfully exposited by an argument due to David Mermin.

Our question is this: Can we prearrange the behavior of our three separated particles so that—each being completely insensitive to which experiment is carried out on the others—the predictions listed above are certain to obtain? To accomplish this, each particle must be physically predetermined to react a certain way to each possible orientation of the magnet. For if it were not, if some irreducible randomness were involved in producing one outcome, then the other particles would have to be sensitive to how the random element came out in order to adjust their behavior properly. Similarly, assuming no physical dependency of any of the particles on the setting of the distant magnets, these predetermined reactions must guarantee the right results no matter how the distant magnets are set. In short, we must somehow fill in the circles in Figure 13 with "up" or "down," specifying how each particle would react to each magnet, in a way that respects the predictions for each of the four global experimental arrangements. There must be an odd number of "ups" along the dashed $X_1X_2X_3$ row and an even number of "ups" along the other three indicated rows.

But it is clear that no specification can meet these requirements. If it could, then adding the total number of ups along all four rows would yield odd + even + even + even = an odd number of ups. However, the entry on each circle would have been counted twice, since each circle lies at the intersection of two rows. Since each circle is counted twice, no matter how we fill in the circles, the total count of all four rows must yield an even number of ups. QED.

The EPR argument shows that if the quantum mechanical predictions are to hold without any spooky-action-at-a-distance (i.e., without any physical sensitivity of any particle to which particular experiment is performed on a distant particle, or to an indeterministic outcome produced by the distant particle), then the reaction of each particle to each possible experimental arrangement it might encounter must be predetermined by its own physical state. But the GHZ argument shows that the set of possible reactions cannot be so predetermined in a way that is insensitive to the distant experimental arrangements. We are stuck with the spooky action-at-a-distance that Einstein so abhorred. This is an example of quantum nonlocality.

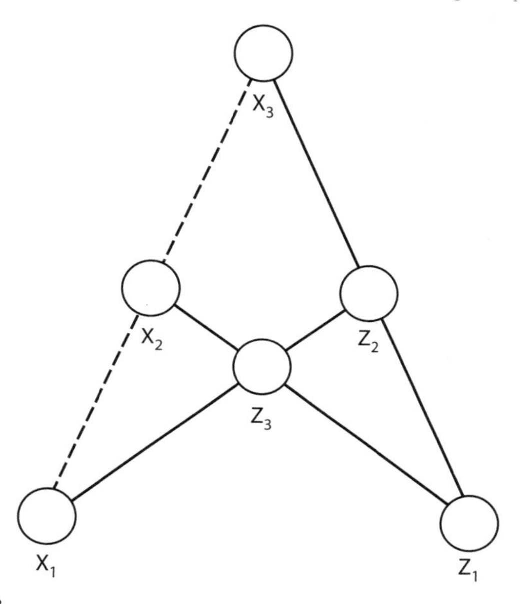

Figure 13

The fact that this nonlocality is required by the predictions of quantum theory was first discovered by John Bell in 1964. Bell's initial arguments did not use the GHZ set-up, with the three separated particles, but rather the EPR set-up for different spin experiments carried out on a pair of particles in the singlet state. Bell began by rehearsing the EPR argument that avoiding nonlocality in this setting requires predetermining the behavior of each particle for all possible orientations of the magnet. But Bell went a step beyond EPR: Instead of considering only the cases where the two magnets are oriented in the same direction (and the particles are certain to display opposite outcomes), he considered the statistical predictions for cases where the magnets on the two sides have different orientations. Bell showed that there are limits on the strength of observed correlations between the sides for different settings of the magnets if one assumes that the outcomes for each possible setting are predetermined and depend only on the orientation of the magnet that the electron passes through. This limit is called *Bell's Inequality*. Quantum theory predicts violations of this limit.

In particular, when the magnets on the two sides are misaligned by 60°, instead of giving different outcomes all the time, they give them 75% of the time. And when they are misaligned by 120°, they give different outcomes 25% of the time. There is no way to preassign dispositions to individual particles that can recover these statistics in the long term.[11] But if the outcomes are not predetermined in this way, then the electron on one side must be sensitive to which experiment is carried out, or what the outcome is, on the other.

Next Step

What we are ultimately interested in is how the physical world can be, given that it produces this sort of observable behavior. Feynman despaired of the very possibility of such a precise physical account of even the two-slit phenomena. As we will see starting in Chapter 4, it is possible to explain how it works. Indeed, there are several different precise physical theories that can explain how it works by postulating an exact physical ontology and precise, mathematically formulated physical laws. But before turning to these alternative theories, we will discuss the thing that Feynman is happy to present: the standard mathematical techniques used by physicists to derive predictions about the sorts of experimental situations described above.

Further Reading

Clear introductions to the foundations of quantum mechanics are few and far between. The best available are noted here.

[11] See Maudlin (2011) for more details.

Albert's (1992) book is aimed at a philosophical audience and includes more discussion of conscious experiential states than the other books mentioned here. Bell (2004; especially chapters 7, 15, 16, 17, 20, 22, 23, and 24) is not a single unified account, and different chapters are pitched at different levels of mathematical sophistication, but his book is still the most profound one we have. Bricmont (2016) is advocating for one particular theory, but he includes a lot of historical material and walks the reader through some of the mathematics. Ghirardi's (2005) book covers more ground than the others and is written at a comfortable level for anyone who has had introductory physics. Norsen (2017) is a textbook on quantum foundations designed for undergraduates, containing homework problems and projects as well as a large dollop of history. Finally, Becker (2018) presents the most comprehensive, reliable, and readable account of the history of the development of quantum theory, drawing in part on the more scholarly and detailed work of Beller (1999).

CHAPTER 2

The Quantum Recipe

WHAT IS PRESENTED in most standard textbooks on quantum physics is not a theory but a recipe: a set of techniques for making predictions. As a predictive mechanism, this recipe is unparalleled in history for both its scope and precision. This requires explanation: What kind of physical structure of the world could give rise to the behavior that is so precisely and accurately predicted by the recipe? Before we can sensibly approach this question, we need to have some detailed understanding of what the recipe itself is. That is the topic of this chapter. We will confine ourselves to the simplest, most unsophisticated version of quantum physics: the nonrelativistic version dealing with spin-1/2 particles, the level of most introductory textbooks. There are much more sophisticated and technically challenging types of quantum physics (e.g., quantum field theory), but the fundamental interpretational and conceptual questions we are interested in can be raised and discussed in this simpler setting. This presentation already has some slightly nonstandard aspects, which will be noted in the section titled "Eigenstates, Eigenvalues, Hermitian Operators, and All That." But for the purpose of understanding our eight experiments, this version of the quantum recipe serves admirably.

SINGLE PARTICLE, NO SPIN

Our first three experiments—the Cathode Ray Tube, the Single Slit, and the Double Slit—concern the behavior of sequences of single particles. Or, more exactly, they concern the formation of single marks or flashes on a screen, produced sequentially, without arrangements for pairs of particles to interact. No electric or magnetic fields are involved beyond the internal workings of the cathode and anode, so the issue of the magnetic properties of the electrons does not come up. How does the recipe work in this setting?

The first step in the recipe requires associating a mathematical object called a *wavefunction* with each electron. The term "wavefunction" is used in different ways in different discussions of quantum theory, but throughout this book, we will be fastidious about its meaning. A wavefunction is a purely mathematical item used for calculational purposes in the quantum recipe. Specifying a wavefunction for a physical system means associating a particular mathematical object with that system, no more and no less. Since a *function* is an abstract mathematical entity—a mapping from one set of objects to another—a wavefunction should, on the face of it, be a mathematical object. We leave aside for now the question of what (if anything) in the physical world this wavefunction represents. Various proposals about this can be distinguished. One might maintain that the wavefunction represents some physical feature of individual physical systems, in which case we will call that feature the *quantum state* of the system. Or one might maintain that the wavefunction only represents the statistical features of collections of physical systems but nothing about single systems. Or one might maintain that the wavefunction represents nothing intrinsic about any physical system at all: Instead it represents some agent's state of information or state of belief about a system. These are incompatible accounts of what the wavefunction of a system represents. But the advocates of these various views will still agree about which mathematical wavefunction ought to be associated with a system in a particular experimental configuration.

If we are dealing with single particles, disregarding their magnetic properties, then the wavefunction takes the mathematical form of a *complex square-integrable function over space, as a function of time*. Let's take these features one at a time.

A complex function over space is a mathematical mapping that assigns a complex number to each spatial location. If we coordinatize the space with the usual Cartesian coordinates (x, y, z), then this function assigns a complex number to each set of coordinate values. Complex numbers are sometimes presented in the form A + Bi, where A and B are both real numbers, and i is the square root of –1. For our purposes, it is more convenient to represent the complex numbers in the form $Re^{i\theta}$, where R is the *amplitude* of the complex number, and θ is its *phase*.

Translating between these two representations of complex numbers is not hard. To go from the amplitude/phase representation to the real part/imaginary part representation, one just has to recall this formula for raising e to an imaginary power:

$$e^{i\theta} = \cos(\theta) + i\sin(\theta).$$

This yields $Re^{i\theta} = R\cos(\theta) + Ri\sin(\theta)$, so $A = R\cos(\theta)$, and $B = R\sin(\theta)$. In the other direction, given A + Bi, we have $R = \sqrt{(A^2 + B^2)}$ and $\theta = \tan^{-1}(B/A)$. (The mathematically inclined might note that this last formula does not have a unique solution, since we can add any multiple of π to θ and still satisfy the equation, measuring θ in radians.) The real part/imaginary part representation is particularly useful when we are adding complex numbers: $(A + Bi) + (A' + B'i) = (A + A') + (B + B')i$. We just add the real and imaginary parts separately. The amplitude/phase representation is useful when multiplying complex numbers: $Re^{i\theta} \times R'e^{i\theta'} = RR'e^{i(\theta + \theta')}$. We multiply the amplitudes and add the phases.

The *complex conjugate* of a complex number is obtained by changing the sign of the imaginary part: The complex conjugate of A + Bi is A − Bi. In the phase/amplitude representation, this amounts to changing the sign of the phase: The complex conjugate of $Re^{i\theta}$ is $Re^{-i\theta}$. The complex conjugate is indicated by an asterisk, so that $(A + Bi)^* = (A - Bi)$, and $(Re^{i\theta})^* = Re^{-i\theta}$. The *absolute square* of a complex number is the number multiplied by its complex conjugate. The reader should be able to verify that the absolute square of (A + Bi) is $A^2 + B^2$, and the absolute square of $Re^{i\theta}$ is R^2. Because of this last identity, the absolute square is also called the *squared amplitude*.

The wavefunction of a single-particle system at a given time is usually symbolized by ψ (or $\psi(x)$) to indicate it is a function of location in space) and the complex conjugate by $\psi^*(x)$. Since $\psi(x)$ assigns a complex number to each spatial location, the product $\psi^*(x)\psi(x)$ assigns a non-negative real number—the *squared amplitude*—to each location. This squared amplitude can be integrated over all space, which amounts to measuring the total volume under the function. A complex function is *square integrable* if this yields a finite number. A square-integrable function can usually be *normalized*, that is, rescaled so that when integrated over the whole space, the result is 1. If the total integral of a square-integrable function is N, then dividing the function by N will normalize it, assuming N is not zero. The wavefunction associated with a system is required to be normalized, for reasons that will soon become apparent.

Given what a wavefunction is, it is obvious that certain mathematical operations on wavefunctions are well defined. For example, we can multiply any wavefunction by a real or complex number by multiplying its value at any spatial point by that number. Two wavefunctions $\psi(x)$ and $\phi(x)$ can be added: the value of $(\psi + \phi)(x)$ at any point is just the sum of $\psi(x)$ and $\phi(x)$ at that point. If $\xi(x) = c\psi(x) + d\phi(x)$, where c and d are complex numbers, then we say that $\xi(x)$ is a *superposition* of $\psi(x)$ and $\phi(x)$. It makes no sense to ask whether a particular wavefunction is a superposition or not, but only whether it is a superposition of some other wavefunctions. Every wavefunction is the superposition of others in many ways.

There is a special function $\mathbf{0}(x)$ which has the value 0 at every spatial point. $\mathbf{0}(x)$ plays the role of the zero of addition of wavefunctions, since $\psi(x) + \mathbf{0}(x)$ always equals $\psi(x)$. Every wavefunction has an additive inverse $-\psi(x)$, whose value at every point is the negative of $\psi(x)$, so that $\psi(x) + (-\psi(x)) = \mathbf{0}(x)$. Due to these features (and a few more), the collection of wavefunctions forms a *complex vector space*, that is, a collection of items that can be added to each other and multiplied by complex numbers. The special function $\mathbf{0}(x)$ is square integrable, but since the integral is zero, it cannot be normalized. If we demand that physical systems be associated with normalizable functions, $\mathbf{0}(x)$ cannot represent any physical system.

So far, all we have asserted is that at any given time, the quantum recipe requires us to associate with each electron a normalized complex-valued function over space. But we have said nothing about 1) *which* such function should be associated with the system 2) how this associated function evolves through time and 3) how predictions are to be derived from the function. These parts of the recipe are not all perfectly precisely formulated. The association of wavefunctions with systems often proceeds via reasoning using classical physics. This procedure might seem conceptually a bit confusing, but if all we want is a functioning recipe, then it is adequate, so long as the instructions are clear enough.

To treat our first three experiments, we need one main rule for associating wavefunctions with systems: If an experimental arrangement, thought of in terms of classical physics, would produce particles with some nearly exact *momentum*, then the right wavefunction to use to represent it is approximately a complex plane wave.

A complex plane wave is a close cousin to the plane waves that occur in water, such as that depicted in Figure 3a. The parallel lines in that figure represent the crests and troughs of the waves. In a similar fashion, one can indicate the points in a complex field that have the same *phase* instead of the same height. In a complex plane wave, these regions of equal phase form parallel lines. And just as the wavelength of a water wave is determined by the distance between successive crests or troughs, the wavelength of a complex plane wave is determined by the distance between successive regions with the same phase.

We will make extensive use of the analogy between complex waves and water waves, but there is one important disanalogy to note. Water waves have only one degree of freedom at a point: the amplitude. Water waves are nothing but regular variations in the height of the water in space. So the squared amplitude of a water wave must also necessarily vary from place to place. But complex waves can have constant amplitude and vary only in phase, as illustrated in Figure 14. This is the sort of complex plane wave associated with a particle with a perfectly exact momentum. In Figure 14, the momentum is horizontally to the right. Since the amplitude of such a complex wave is constant, the squared amplitude is constant, rather than varying from place to place.

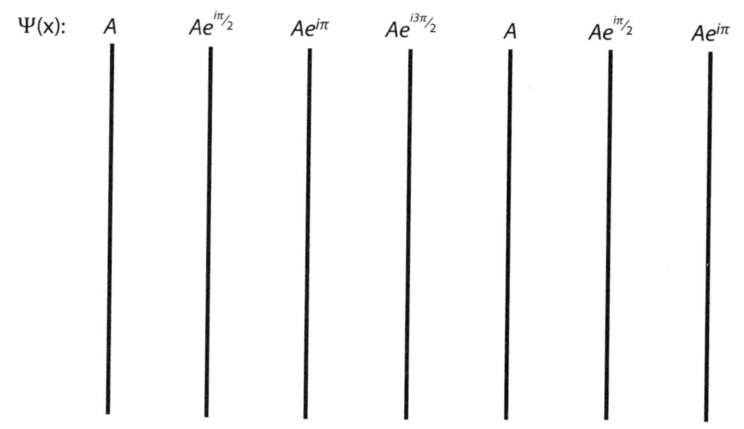

Figure 14

Classically, a cathode ray tube would be regarded as a device for producing electrons with a (fairly) constant momentum collimated in a beam. Each electron that comes off the cathode is accelerated the same amount by the voltage potential between the cathode and the anode. Those electrons that pass through the aperture in the anode therefore have (approximately) the same momentum, in the direction from the cathode to the anode. Our treatment of the first four experiments will use a complex plane wave to represent these electrons.

We need to be more mathematically exact than this, because we must precisely specify the wavelength of our plane wave. Let us appeal to a rule that connects the classical treatment of this experimental situation with the quantum treatment. The connection was provided by Louis de Broglie, who formulated this rule for ascribing a wavelength to particles with a definite momentum:

$$\lambda = h/\mathrm{p},$$

where λ is the wavelength, h is Planck's constant, and p is the momentum of the particle. (Classically, the momentum is m**v**, where m is the mass and **v** the velocity.) Using de Broglie's formula, we can associate a definite sort of wavefunction with a particle that would classically have been prepared to have a definite momentum: a complex plane wave whose wavelength is h/p, where the direction of motion is orthogonal to the regions of constant phase.

To make predictions, we also need a rule for how these complex waves evolve in time. We now consider the wavefunction as a function of both space and time $\psi(x, t)$. The rule for its change in time is provided by *Schrödinger's equation*. In the nonrelativistic context, the abstract form of the equation is

$$i\hbar \frac{\partial \psi(x,t)}{\partial t} = \widehat{H}\psi(x, t),$$

where i is the square root of -1; $\hbar$ is Planck's constant divided by 2π; $\frac{\partial \psi(x,t)}{\partial t}$ quantifies how $\psi(x, t)$ changes over an infinitessimal amount of time; and $\widehat{H}$ is the *Hamiltonian operator*. An *operator* is a function that maps one wavefunction to another, and the "hat" over the H is used to indicate an operator. This general abstract form of the Schrödinger equation has no real content until the operator $\widehat{H}$ is specified. Here, once again, an appeal is made to classical physics.

The Hamiltonian of a system is related to the classical notion of the total energy of the system. For a classical particle, that energy has two sources: the kinetic energy ("energy of motion") and the potential energy, deriving from things like gravitational or electrical or magnetic potentials. The classical kinetic energy of a particle of mass m is $\frac{1}{2}\mathrm{m}v^2$. The potential energy depends on which sorts of potentials are present, but they are usually some function of the mass, electric charge, and so forth of the particles and their distance from one another. So in a classical setting, the total energy of a particle of mass m is usually given by the equation

$$\mathrm{E} = (\tfrac{1}{2})\mathrm{m}v^2 + \mathrm{V}(x),$$

with $\mathrm{V}(x)$ being some function that depends on the experimental arrangement.

To specify Schrödinger's equation, we need to create a Hamiltonian operator that is modeled on the equation above. For the moment, we will pretend that there is only one spatial dimension, x, that the particle can move in. The technical trick is to "put the hats on," that is, to replace variables that represent physical quantities in the classical equation (e.g., the velocity v and the position x of the particle), with operators. In the case of the position x, this replacement is very simple: the operator $\widehat{x}$ simply multiplies a wavefunction by the variable x, yielding a new wavefunction. Replacing v is slightly more complicated. Since the classical momentum p is mv, we can write $v = \frac{p}{m}$, and rewrite the kinetic energy as $\frac{1}{2m}p^2$. And now we use a basic instruction of the recipe: when putting a hat on p (i.e., changing the classical momentum into an operator), the operator to use is $-i\hbar\frac{\partial}{\partial x}$, that is, $-i\hbar$ times the spatial derivative of the wavefunction. The derivative operator $\frac{\partial}{\partial x}$ yields the *slope* of a function as one moves in the x-direction.

Following the instructions of the quantum recipe, we convert the classical formula for the total energy of a particle $E = \frac{1}{2} m\mathbf{v}^2 + V(\mathbf{x})$ into a definition of the Hamiltonian operator:

$$\widehat{H} = \frac{1}{2m}\hat{p}^2 + V(\hat{x}) = \frac{1}{2m}(-i\hbar\partial/\partial x)^2 + V(\hat{x}) = \frac{-\hbar^2}{2m}\frac{\partial^2}{\partial x} + V(\hat{x}).$$

To deal with a particle that can move in all three spatial directions, we replace x with a three-dimensional position vector $\mathbf{r}$ and the spatial derivative in the x-direction with a three-dimensional spatial derivative ∇. So a standard quantum physics textbook will present the Hamiltonian as

$$\widehat{H} = \frac{-\hbar^2}{2m}\nabla^2 + V(\hat{\mathbf{r}})$$

and Schrödinger's equation as

$$i\hbar\frac{\partial\psi(\mathbf{r},t)}{\partial t} = \left[\frac{-\hbar^2}{2m}\nabla^2 + V(\hat{\mathbf{r}})\right]\psi(\mathbf{r},t).$$

Whew![1]

[1]The rule of "putting the hats on"—replacing classical variables with operators—works unambiguously in many circumstances but not all. Here is a case where it is ambiguous. Classically, the expressions xp and px are mathematically equivalent: the product of position and momentum, But the operators $\hat{x}\hat{p}$ and $\hat{p}\hat{x}$ are not the same. Consider applying them to the function $f(x)$. $\hat{x}\hat{p}f(x) = \hat{x}\frac{\partial}{\partial x}f(x) = x\frac{\partial}{\partial x}f(x)$, but $\hat{p}\hat{x}f(x) = \hat{p}xf(x) = \frac{\partial}{\partial x}x\, f(x) = f(x) + x\frac{\partial}{\partial x}f(x)$. In technical

The last couple of pages contain a lot of equations, which may be daunting to some readers. Don't worry. We are not going to be calculating with these equations, but it important to see 1) how the quantum recipe arrives at precise equations to use and 2) the way that appeal to classical physics motivates the choice of both the equation for how the wavefunction evolves in time and the choice of an initial wavefunction to associate with a system. Physicists who have extensive training in classical physics have no difficulty following the recipe in familiar experimental situations.

To get a sense of just how familiar these equations are, a quick comparison suffices. If there are no relevant potentials, so $V(\hat{r})=0$ (e.g., in empty space), Schrödinger's equation becomes

$$i\hbar\frac{\partial\psi(\mathbf{r},t)}{\partial t} = \frac{-\hbar^2}{2m}\nabla^2\psi(\mathbf{r},t).$$

The classical wave equation, which governs water waves and waves in elastic materials and waves in vibrating strings is

$$\frac{\partial^2 u(x,t)}{\partial t^2} = c^2\nabla^2 \mathrm{u}(x,t).$$

The wave equation uses a second derivative in time, and $\mathrm{u}(x,t)$ is a real function rather than a complex function: $\mathrm{u}(x,t)$ represents the amplitude of the wave at a given place and time. The classical heat equation, which describes how the temperature distribution of, for example, a bar of metal changes is

$$\frac{\partial u(x,t)}{\partial t} = \alpha\nabla^2 \mathrm{u}(x,t),$$

which looks even more like Schrödinger's equation, except again $\mathrm{u}(x,t)$ is a real function, specifying the temperature. And since ψ for a single particle is itself a function in physical space, we could write it as $\psi(x,t)$ or $\psi(\mathbf{r},t)$. So aside from the introduction of complex numbers, the mathematics so far is quite familiar from classical physics.

---

terminology, $\hat{p}$ and $\hat{x}$ do not commute. This particular issue does not come up in our experiments.

Indeed, the reader might well at this point wonder why the quantum recipe doesn't just yield predictions that are extremely similar to those of classical physics. The mathematical formalism, aside from using complex numbers, is in some cases identical. So how can it lead to predictions that are at all surprising or shocking to classically honed expectations?

But the quantum recipe has three distinct parts: rules for assigning a wavefunction to a system, rules (in this case, Schrödinger's equation) for how that wavefunction evolves through time, and rules for *extracting predictions about observable phenomena* from the wavefunction. So far, we have said nothing at all about this last, critical step.

Before confronting the last step, let's see how the first two steps play out for Experiment 1. Thinking classically and initially regarding the cathode rays as classical particles with mass m_e (the mass of the electron) and charge q_e (the charge of the electron), we can calculate from the voltage between the cathode and anode how fast the electron should be traveling when it reaches the anode and hence what its momentum $m_e\mathbf{v}$ would be. We therefore associate with the cathode ray a complex plane wave of wavelength $h/(m_e v)$. Since we are only interested in cathode rays that pass through the aperture in the anode, we initially restrict this plane wave to the region just in front of the aperture. And since there are no further potentials, we can use Schrödinger's equation to calculate how the wavefunction will evolve in time. Assuming that the aperture in the anode is much, much bigger than the wavelength, the plane waves will essentially just progress further ahead in the same direction until they reach the screen (as in Figure 3a). Then what?

If one took the analogy to classical waves even further, one would expect the wavefunction to interact with the screen. And since the wavefunction is spread evenly over an aperture-shaped area of the screen, one might expect that whatever effect that interaction might engender, it would also be spread evenly across that area of the screen. And if one were dealing with a fairly strong beam of cathode rays, this would correspond to what one sees: a uniformly glowing patch of screen roughly the same size and shape as the aperture in the anode (and a little fuzzy around the edges). But we have not been trying to make predictions about electron or cathode ray beams. We have been trying to make predictions about individual electrons. And in this case, which corresponds to turning the intensity of the cathode ray tube down, we know that we do not see anything uniformly spread across that area of the screen. Rather, we see individual, localized flashes or spots. Each flash occurs in the area that the wavefunction reaches, but each is also localized in a very small part of that region. If we let a lot of these spots accumulate over time, we find that they are roughly uniformly distributed over the region. But each particular flash is not.

There is nothing in the first two steps of the recipe that would suggest this sort of behavior. But again, since the recipe is just a recipe and not a theory, it is hard to see how it could reasonably suggest anything at all. What is needed, as I have been insisting, is just a rule: a set of instructions about how to use the wavefunction for making predictions. The requisite rule was originally suggested by Max Born, and hence is known as Born's Rule. Born suggested using the wavefunction to define a *probability measure*, and then using that probability measure to make predictions about where an individual flash or spot will occur.

A probability measure, in the purely mathematical sense of that term, must have certain formal properties. A probability measure assigns numbers to a collection of possible outcomes. This collection is called a *sample space*. The measure assigns a real number between 0 and 1 (inclusive) to the *measureable subsets* of the sample space. To be a probability measure, the measure must assign 1 to the whole set and must be *countably additive*. The latter condition means that if the measure assigns the values p_i to (possibly countably infinite) disjoint subsets σ_I of the sample space, then it must assign the sum of the p_i to the union of those subsets.

So if Born's rule is to tell us how to use the wavefunction to make predictions by defining a probability measure, it must specify both what the outcomes are and what the measure is. In our experiments, an "outcome" is going to be a flash or spot occurring on the screen at a particular location, so we can think of the probability measure as defined over the various locations on the screen. And since the wavefunction itself is (in this case) defined over physical space, we might be tempted to take the value of the wavefunction at the various locations on the screen to be the measure. But as is obvious, the wavefunction itself is not of the right mathematical form to play this role: It assigns complex numbers to locations in space, not real numbers between 0 and 1. However, if we take the squared amplitude of the wavefunction, we get just the right kind of thing to be a probability measure or, more precisely, a probability density. It assigns to every location a non-negative real number, and if we integrate these numbers over any region of the screen, we get a real number between 0 and 1. If we integrate the density over the entire screen (the whole "outcome space"), then we get exactly 1, assuming that the wavefunction has been normalized. That is why, as was mentioned above, wavefunctions are usually required to be normalized, so their squared amplitude can serve as a probability measure.

With Born's rule in place, the recipe is complete. To predict how likely it is that a flash will occur in any given region of the screen, compute the wavefunction in that region, take its squared amplitude, and integrate the result over the region. Where the squared amplitude is high, there is more likely to be a flash; where it is low less likely; and where the amplitude is zero, there is zero chance. Because of this connection between the wavefunction and probabilities, wavefunctions are sometimes called "probability waves," but this term is inaccurate. It is the squared amplitude of the wavefunction that yields a probability (or a probability density). A better terminology is "probability amplitude."

It should be emphasized again that nothing in the first two steps of the recipe implies, or even suggests, that the squared amplitude of the wavefunction should be used to define a probability. Certainly nothing in the classical analogs—water waves and the heat equation—has anything probabilistic about it. Born's rule comes out of nowhere, and it injects probabilistic considerations into the physics without warning. Nonetheless, the resulting recipe works with spectacular accuracy.

If we have a beam of electrons, each with the same wavefunction, then the recipe treats them as probabilistically independent: Where one flash occurs gives no information about where any other will occur. So if we let many, many flashes occur, it becomes overwhelmingly likely that their distribution reflects the probabilities. For example, if the probability measure assigns 0.4 to some region on the screen, then very nearly 40% of the flashes will occur there. If the squared amplitude of the wavefunction is constant over some part of the screen, then a beam of electrons should produce a glow of constant brightness; and if the squared amplitude varies, so should the glow. In this way, our theory produces definite predictions about the overall pattern produced by many electrons over many runs.

But since Born's rule assigns a probability for results of experiments with single electrons, we get the correct results there as well. On a single run, there will be a single flash in a single place. The 0.4 calculated is the probability for that single flash to occur in the given region.

There is still one important puzzle about Born's rule: Under what circumstances, exactly, is one allowed to use it? What are the sorts of outcomes to which these probabilities can be attached?

The answer is usually phrased in terms of measurement. Use Born's rule, we are told, when a measurement is made on an electron, and use it to assign probabilities to the possible outcomes of that measurement (which should be various possible positions or locations for a position measurement). But as to what, precisely, a "measurement" is, when one occurs, and what exactly is measured, Born's rule is silent. Such judgments about when to use the rule are left to the discretion of the physicist. In our examples, flashes on the screen are taken to obviously be position measurements and the flashes themselves as "outcomes." So in these particular cases, there is no practical impediment to following the recipe, even though in other situations it might not be obvious how to proceed.

What about other sorts of measurements? I have discussed Born's Rule so far as a recipe for extracting predictions about position measurements from the quantum recipe, but, the objectors may insist, this unjustifiably privileges position measurements over other sorts of experiments, such as momentum measurements or energy measurements or spin measurements. Born's Rule, they may say, covers all these sorts of experiments, not just position measurements. My presentation of the rule biases the account to favor positions of things over momentum or energy or spin.

John Bell addressed this issue directly and forcefully:

> The second moral is that in physics the only observations we must consider are position observations, if only the positions of instrument pointers. . . . If you make axioms, rather than definitions and theorems, about the 'measurement' of anything else, then you commit redundancy and risk inconsistency.[2]

Let's unpack what Bell means. Suppose there is an experiment called a "momentum measurement," or an "energy measurement," or a "spin measurement" on an electron, and one wants to make a prediction about how it will come out. And suppose that, for example, a momentum version of Born's Rule is proposed that allows one to derive probabilities for the possible outcomes of such an experiment. To be concrete, suppose one derives a 50% chance that the momentum of the electron will be recorded as +2 and a 50% chance it will be recorded as –2. Some actual experiment in the lab is carried out to check this prediction. But now let's step back and treat the entire lab set-up—the original electron together with the laboratory apparatus—as a physical system (which it surely is!). We want physics to provide a prediction for the position of the ink on the computer output at the end of the experiment. This is, of course, asking about the positions of things, so the position version of Born's Rule can be invoked.

Bell's point is this. Either the application of the position version of Born's Rule to the physics of the whole laboratory set-up yields a 50% chance that the spatial pattern "+2" is printed and a 50% chance that the spatial pattern "–2" is printed, or it yields

[2] Bell (2004), p. 166.

something else. If it yields something else, then the momentum version of Born's Rule is not actually providing predictions for the data produced by the experiment, so our theory has become inconsistent. That is, if we conceptualize it as a momentum measurement and use the momentum version of Born's Rule, we get one set of predictions, but conceptualized as just a physical interaction that ends up with ink on paper, we get a different set of predictions. But what ends up on the paper is independent of how we conceptualize the situation! The momentum-rule prediction for the outcome of the experiment therefore had better not differ from the position-rule prediction when applied to the whole apparatus. In contrast, if the position-measurement prediction for the ink always agrees with the momentum-version prediction, then the momentum version of the rule is dispensable: We are actually comparing the predictions of the theory against the printed outputs of the lab equipment. If the theory gets that right, then it gets "the experimental data" right. In short, we need an empirical theory to get positions right, and if the theory does that, then it fulfills all requirements for empirical accuracy.

In any case, the particular experiments we want to account for are naturally understood as culminating in a position measurement. When we measure spin using a Stern-Gerlach apparatus, the outcome is registered in the deflection of a particle up or down, that is, in the position of the particle. So our discussions can proceed smoothly while restricting the application of Born's rule to position measurements.

What does our predictive recipe predict? For Experiment 1, as we have seen, the experimental apparatus would classically produce electrons with a fairly precise momentum, so we choose a wavefunction with the corresponding wavelength and a constant amplitude to represent the electron. The spatial structure of the wavefunction is a little vague. It should start out roughly the shape of the aperture in the anode, but we let the amplitude drop gently to zero at the edges (gently with respect to the scale of the wavelength). Schrödinger's equation then propagates the wavefunction forward in time. The ∇^2 term in Schrödinger's equation is sensitive to how quickly the slope of the wavefunction changes in space, and the equation implies that the more quickly it changes in some spatial direction, the more rapidly the wavefunction expands in that direction. So by making the edges of the wavefunction drop gently at the sides, we ensure that the wavefunction itself does not spread much in those directions. The wavefunction propagates forward until it reaches the screen, and the squared amplitude is constant across a region of the screen roughly the shape of the aperture, dropping to zero outside that region. By Born's rule, we predict an equal chance for the flash to occur in any equal-sized parts of that region, with no chance for a flash to occur outside. This is exactly what we observe when we run Experiment 1.

For Experiment 2, we add a second barrier with a much smaller hole or slit between the anode and the screen, and we attend only to flashes at the screen (i.e., to electrons that "make it through" the slit; Figure 3b). The overall situation is quite similar to Experiment 1, except that as we make the slit progressively smaller, we eventually reach a point where the wavefunction just beyond the slit cannot taper off gently to zero on the scale of the wavelength. If the slit itself is only as wide as the wavelength, for example, then the slope of the wavefunction must vary fairly rapidly in the region just beyond the slit. And as we have seen, Schrödinger's equation implies that a wavefunction whose slope varies greatly in a spatial direction also spreads rapidly in that direction. Having been confined to a very narrow slit, then, the wavefunction subsequently spreads out, resembling a circular wave rather than a plane wave. The narrower the slit is (relative to the wavelength), the greater the spread will be. This yields exactly the behavior observed in Experiment 2.

Another detail fits as well. If we increase the voltage between the cathode and the anode, then classically, we expect the electrons to have a higher momentum. But in the recipe, higher momentum means shorter wavelength. So a slit that causes a lot of diffraction at a low voltage should produce less and less diffraction at higher voltage, as the wavelength becomes small compared to the slit. We see exactly this behavior.

Finally, we come to Experiment 3, the Double Slit. Predicting the outcome of the Double Slit using the recipe is almost child's play once one important mathematical fact is remarked. Schrödinger's equation has the very convenient mathematical feature called *linearity*. This means that just as two wavefunctions ψ and ϕ can be superposed to form a third wavefunction, so, too, the solutions to Schrödinger's equation generated by ψ and ϕ can be added to yield the solution generated by $\psi + \phi$. Here's how to parlay that feature into a prediction for the Double Slit.

We know that with only one slit open, the wavefunction just beyond the slit spreads out in a semicircular pattern: Call this wavefunction $\psi(\mathbf{r}, t)$ (Figure 15a). With the other slit open, we get the same pattern but moved slightly in space (Figure 15b). Call this $\phi(\mathbf{r}, t)$. With both slits open, the wavefunction at $t = 0$, the moment it just passes the slits, is an equal superposition of $\psi(\mathbf{r}, 0)$ and $\phi(\mathbf{r}, 0)$: a wavefunction with one lump just in front of one slit and another equal-size lump just in front of the other. The linearity of Schrödinger's equation then implies that the solution at all times is just the equally weighted sum of the solution for $\psi(\mathbf{r}, 0)$ alone and the solution for $\phi(\mathbf{r}, 0)$ alone. But both $\psi(\mathbf{r}, t)$ and $\phi(\mathbf{r}, t)$ are complex functions, and so their sum can exhibit interference (Figure 15c). In these images, we are using the apparent height of the wave as a proxy for the value of the phase of the wavefunction, as illustrated in Figure 14.

Indeed, the interference manifests itself just as it does for water waves. The two parts of the wavefunction at $t = 0$, the parts in front of the two slits, have equal magnitude and phase, because the plane wave that hit the barrier had equal magnitude and phase in those locations. At any point on the screen where the difference in the distances to the two slits is a multiple of the wavelength, the two superposing waves arrive with the same phase, and the resulting wavefunction has twice the amplitude of each. But if the difference of the distances is a half wavelength (or 3/2, or 5/2, etc.), then the two superposing wavefunctions have opposite phase and equal amplitude. Added together at that point they cancel out ($e^{i(\theta + \pi)} = -e^{i\theta}$), leaving the wavefunction with zero amplitude. By Born's rule, a flash has no chance to occur there.

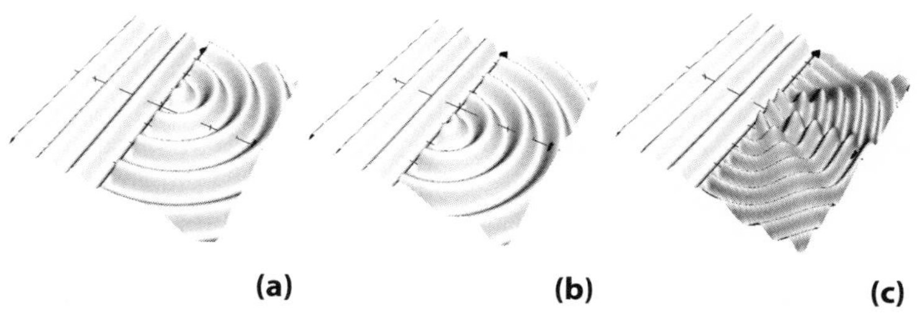

Figure 15

The alternating regions of high probability and zero probability yield the interference fringes as many flashes accumulate. The arrangement of the interference bands follows from the wavelength of the wavefunction and the geometry of the slits and the screen.

The predictive recipe handles our first three experiments with relative ease. But why do the interference bands disappear when the monitoring proton is added?

Multiple Interacting Particles, No Spin

Experiment 4 adds a new wrinkle: we have to take account of both the electron and proton on each run. At a mathematical level, this turns out to be simpler than one might have guessed.

One's first thought is that for an experiment involving two particles, the recipe should employ two wavefunctions, one for each particle. This is not what is done, and therein lies the key to the most surprising predictions derived from the recipe. For a single particle, the wavefunction is a complex function over physical space. But for multiple particles, the wavefunction is a complex function over the *configuration space* of the system.

Once again, we appeal to a classical picture of point particles to explain this notion. The *configuration* of a collection of point particles is a specification of where each of the particles is. For a single particle, this amounts to the specification of where that single particle is—the specification of a point in space—so for a single particle system, the configuration space is isomorphic to physical space. We might represent the single-particle space by three coordinates (x, y, z). But if the system has an electron and a proton, specifying where they both are requires six coordinates: $(x_e, y_e, z_e, x_p, y_p, z_p)$. So the configuration space of two distinguishable particles is a six-dimensional space, whose points are represented by a six-tuple of real numbers. This space is called R^6. The configuration space of three distinguishable particles is R^9, and in general the configuration space of N distinguishable particles is R^{3N}. The wavefunction associated with such a system assigns a complex number to each point in the configuration space.[3]

There is one particularly simple way to generate a wavefunction over a two-particle configuration space. Suppose one has a single-particle wavefunction for an electron $\phi(x_e, y_e, z_e)$ and a single-particle wavefunction for a proton $\xi(x_p, y_p, z_p)$. Each of these wavefunctions assigns a complex number to each set of values for its coordinates. So one can just multiply these wavefunctions together to get a wavefunction over the configuration space of the two particles: $\psi(x_e, y_e, z_e, x_p, y_p, z_p) = \phi(x_e, y_e, z_e)\, \xi(x_p, y_p, z_p)$. This is called a *product state* of the two-particle system.

Quantum theory would not be nearly so interesting if all multiparticle states were product states: In a product state, the behavior of one particle is uncorrelated with the behavior of the other. As a purely mathematical matter, product states are very scarce in the set of all wavefunctions over our two-particle system. Most wavefunctions cannot be expressed as the product of a wavefunction for the electron with a wavefunction for the proton. Any wavefunction that cannot be so expressed is called an *entangled* state. Erwin Schrödinger introduced the term "entanglement"

[3] What if the particles are not distinguishable in any way? Then specifying the configuration of an N-particle system is just a matter of indicating a set of N points in physical space, since there is no further fact about which particle goes in which location. If the physical space is represented by R^3, then the configuration space for N particles corresponds to all sets of N points in R^3. In Dürr et al. (2006), this space is called ${}^N R^3$. The space ${}^N R^3$ is mathematically different from R^{3N}, so a complex function over ${}^N R^3$ is mathematically different from a complex function over R^{3N}.

("Verschränkung" in German) in his famous cat paper, and he wrote in 1935:

> When two systems, of which we know the states by their respective representatives [i.e. wavefunctions], enter into temporary physical interaction due to known forces between them, and when after a time of mutual influence the systems separate again, then they can no longer be described in the same way as before, viz. by endowing each of them with a representative of its own. I would not call that *one* but rather *the* characteristic trait of quantum mechanics, the one that enforces its entire departure from classical lines of thought. By the interaction the two representatives (or ψ-functions) have become entangled.[4]

Schrödinger envisages a situation that starts out in a product state and evolves, via an interaction, into an entangled state. That is precisely what happens in Experiment 4.

Before embarking on the analysis, let's pause to recall that Feynman described the Double Slit experiment as "a phenomenon which is impossible, *absolutely* impossible, to explain in any classical way, and which has in it the heart of quantum mechanics. In reality, it contains the *only* mystery." But the Double Slit experiment makes no use of the entanglement of systems, and it is the simple Double Slit experiment that Feynman thinks cannot be modeled using classical probability theory. Yet 28 years earlier, Schrödinger had declared that entanglement was "the characteristic trait of quantum mechanics, the one that enforces its entire departure from classical lines of thought." Curiously, the effect of entanglement in the Double Slit with Monitoring is to *make the interference pattern in the regular Double Slit go away*. Feynman noted this phenomenon, but never invoked entanglement to explain it.

All the tools we need to make predictions for this experiment are already in place. The electron effectively moves in only two dimensions: the plane in which the wavefunction propagates. And

[4] Schrödinger (1935), p. 555.

the proton moves in only one dimension: up or down in the cavity depicted in Figure 6. So the relevant configuration space for this experiment is three-dimensional, and we can draw pictures of it.

Consider first what would happen if the lower slit were closed. If the electron gets through, the resulting wavefunction spreads out semicircularly from the upper slit, and the proton wavefunction moves entirely into the upper part of the cavity. We know this, because all the flashes associated with the proton occur there. If we block the upper slit, the electron part of the wavefunction spreads out from the lower slit, and the proton part moves to the lower part of the cavity. This behavior of the proton part is produced by an interaction potential in the Hamiltonian of the system. Without such a potential (which is a function of the distance between the electron coordinate and the proton coordinate), the behavior of the proton would be uncorrelated to that of the electron. The wavefunctions for these two situations are depicted in configuration space in Figures 16a and 16b. Note that in these figures, the upward *z* direction now indicates the position of the proton in its cavity. The "empty" parts of the diagram indicate regions of configuration space where the wavefunction is close to zero.

What if both slits are open? By the superposition principle, the wavefunction just beyond the slits is the sum of the wavefunctions with each slit open. And by the linearity of Schrödinger's equation, the evolution of that superposition is the superposition of their individual evolutions. So we get the wavefunction with both slits open by superposing Figures 16a and 16b, yielding Figure 16c. (The "waviness" of the wavefunction in these diagrams represents the complex phase of the wavefunction, while the upward and downward branching represents the position of the proton as it moves up or down.)

Mathematically, this is the situation. If the lower slit closed, then the system evolves into the product state $\phi_{\text{upper}}(x_e, y_e, z_e)$ $\xi_{\text{up}}(x_p, y_p, z_p)$, where $\phi_{\text{upper}}(x_e, y_e, z_e)$ is the dispersing wavefunction one normally gets for an electron going through the upper slit, and $\xi_{\text{up}}(x_p, y_p, z_p)$ is the wavefunction of a proton in the upper part of the cavity. If the upper slit is closed, then the system evolves

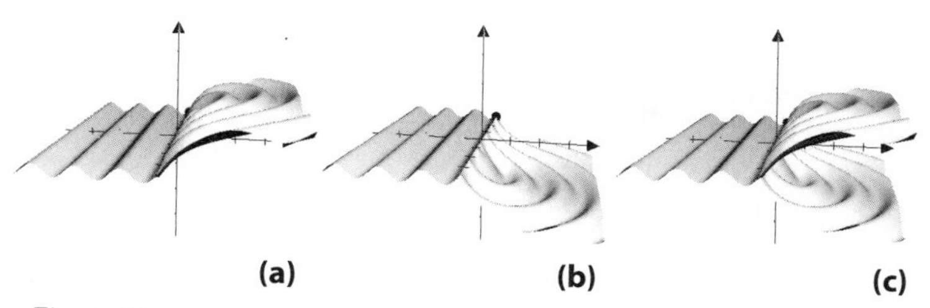

Figure 16

into the state $\phi_{\text{lower}}(x_e, y_e, z_e)\, \xi_{\text{down}}(x_p, y_p, z_p)$, where $\phi_{\text{lower}}(x_e, y_e, z_e)$ is a wavefunction spreading out from the lower slit, and $\xi_{\text{down}}(x_p, y_p, z_p)$ is the wavefunction of a proton in the lower part of the cavity. Therefore, by the linearity of the Schrödinger equation, the initial state

$$\left(\frac{1}{\sqrt{2}}\phi_{upper}(x_e, y_e, z_e) + \frac{1}{\sqrt{2}}\phi_{lower}(x_e, y_e, z_e)\right)\xi_{middle}(x_p, y_p, z_p),$$

which represents the electron as equally likely to go through each of the two slits and the proton initially in the central location, will evolve into the entangled state

$$\frac{1}{\sqrt{2}}\phi_{upper}(x_e, y_e, z_e)\xi_{up}(x_p, y_p, z_p) + \frac{1}{\sqrt{2}}\phi_{lower}(x_e, y_e, z_e)\xi_{down}(x_p, y_p, z_p).$$

This entangled wavefunction is represented in Figure 16c.

But in Figure 16c, there is no longer any interference between the part of the wavefunction associated with the upper slit and the part associated with the lower: these two pieces of the wavefunction have become separated in configuration space because of the proton's contribution to the configuration. The squared amplitude at the screen of the wavefunction in Figure 16c is just the sum of the squared amplitudes of Figures 16a and 16b at the screen. The Schrödinger evolution with both slits open entangles the electron and proton parts, even though the initial wavefunction is a product state. This separation of the wavefunction in configuration space has nothing to do with the existence of any observation or visible result: the separation occurs via strict Schrödinger evolution due to the strong coupling of the electron and proton. In the simple Double Slit experiment, the final wavefunction with one slit open overlaps the final wavefunction with the other slit open in configuration space. When we superpose the solutions to get the solution with both slits open, the two wavefunctions interfere. But when the proton and coupling is added, the two solutions no longer overlap in configuration space, so there is no interference.

The disappearance of interference bands in this experiment is due to *decoherence*. Roughly speaking, if we express a wavefunction $\psi(x, t)$ as the superposition $\alpha\phi(x, t) + e^{i\theta}\beta\chi(x, t)$, then $\psi(x, t)$ has decohered just in case all present and future Born rule predictions are unaffected by the value of θ: $|\psi(x, t)|^2 \approx |\alpha|^2| \phi (x, t)|^2 + |\beta|^2| \chi (x, t)|^2$. In such a case, there is no interference between $\alpha\phi(x, t)$ and $\beta\chi(x, t)$. The electron wavefunction, which displays interference when there is no monitoring proton, loses its interference when entangled in the right way with the proton. The more a given system interacts with other systems, the more entangled it becomes, and the more it tends to decohere. Experiments done on such a decohered system exhibit no interference. So if one takes interference to be the calling card of quantum theory, entanglement and decoherence make the world appear *less* quantum mechanical. But since the cause of the decoherence is entanglement, by Schrödinger's lights, the observable interference disappears because the world is *more* quantum mechanical!

Entanglement and the consequent decoherence explain why we do not encounter quantum interference effects in everyday life. Avoiding decoherence requires severely limiting the interactions a system has with its environment (and even with parts of itself). Such isolation usually requires carefully prepared laboratory conditions.

If we slowly decrease the interaction potential between the electron and the proton, the two separated branches of the wavefunction in Figure 16c will slowly merge, and the interference bands will slowly reappear. When the potential is completely turned off, the proton and electron will no longer entangle at all, and the original double slit interference pattern reemerges.

When Feynman comments on the monitored double-slit experiment, he reports the moral in striking terms: "We must conclude that *when we look at the electrons* the distribution of them on the screen is different than when we do not look" and "*If the electrons are not seen, we have interference!*"[5] This talk of "looking" and "seeing" suggests that the state of the interference bands can only be accounted for if we can characterize acts of observation, and maybe even observers. Bell's worries about whether a single-celled organism can make an observation lurk nearby. So we should be relieved to find that accurate prediction of the phenomenon requires no such thing. All we need to produce the right prediction is the right interaction potential between the electron and the proton in the Hamiltonian of the system. The proton need not, in any interesting sense, be an "observer" or "see" anything. Indeed, the proton need not, itself, produce any flash or other observable phenomenon: the analysis goes through just the same if there is no phosphorescent screen in the cavity. All that is needed is that the wavefunction divide into two non-overlapping regions in configuration space.

Single Particle with Spin

The experiments using Stern-Gerlach magnets demonstrate that another physical degree of freedom of our electrons has to be mathematically represented in the wavefunction. This is accomplished by using wavefunctions that associate something more complicated than a single complex number with each point in configuration space. For spin-1/2 particles, these wavefunctions associate pairs of complex numbers called *spinors* with each point. More precisely, a spinor is a matrix of two complex numbers $\left[\begin{smallmatrix}\alpha\\ \beta\end{smallmatrix}\right]$ such that $|\alpha|^2 + |\beta|^2 = 1$. $|\alpha|^2$ is the absolute square of the complex number α.

By what rules do we associate a particular spinor with an electron? First, we arbitrarily choose a direction in space, which we

[5] Feynman (1975), Volume 1, Section 37-6.

will call the z-direction. We know that we can prepare a beam of electrons so that if it is passed through a Stern-Gerlach magnet oriented in the z-direction, the entire beam will be deflected "up" (i.e., toward the pointy North Pole of the magnet). Such a z-spin up electron is associated with the spinor $\begin{bmatrix}1\\0\end{bmatrix}$. A z-spin down electron is then associated with $\begin{bmatrix}0\\1\end{bmatrix}$.

Note that every possible spinor can now be expressed in terms of these two: $\begin{bmatrix}\alpha\\\beta\end{bmatrix} = \alpha\begin{bmatrix}1\\0\end{bmatrix} + \beta\begin{bmatrix}0\\1\end{bmatrix}$, or, as we will write, $\alpha|z\uparrow> + \beta|z\downarrow>$.

The normalization of the spinor such that $|\alpha|^2 + |\beta|^2 = 1$ allows us to derive a probability measure from it, just as the normalization of the spinless wavefunction allowed us to derive a probability density. The way this probability is used to make predictions is straightforward: given an arbitrary spinor $\begin{bmatrix}\alpha\\\beta\end{bmatrix}$, if we pass an electron associated with that spinor through a Stern-Gerlach magnet oriented in the z-direction, the probability that it is deflected up is $|\alpha|^2$, and the probability it is deflected down is $|\beta|^2$.

So far, the spinor might appear to just be a mathematical device for storing information about the probabilities that an electron will be deflected one way or the other by a z-oriented Stern-Gerlach magnet. But it is much more than that. For the spinor can be used to make predictions about how the electron will behave if passed through a Stern-Gerlach magnet oriented in *any* direction. Here's how that works.

Just as we associate the spinor $\begin{bmatrix}1\\0\end{bmatrix}$ with "up-spin in the z-direction" and $\begin{bmatrix}0\\1\end{bmatrix}$ with "down-spin in the z-direction", so we can associate $\frac{1}{\sqrt{2}}\begin{bmatrix}1\\1\end{bmatrix}$ with "up-spin in the x-direction" and $\frac{1}{\sqrt{2}}\begin{bmatrix}1\\-1\end{bmatrix}$ with "down-spin in the x-direction." And just as an arbitrary spinor $\begin{bmatrix}\alpha\\\beta\end{bmatrix}$ can be expressed as $\alpha\begin{bmatrix}1\\0\end{bmatrix} + \beta\begin{bmatrix}0\\1\end{bmatrix}$, so, too, can an arbitrary spinor be expressed in terms of $\frac{1}{\sqrt{2}}\begin{bmatrix}1\\1\end{bmatrix}$ and $\frac{1}{\sqrt{2}}\begin{bmatrix}1\\-1\end{bmatrix}$. Specifically, $\begin{bmatrix}\alpha\\\beta\end{bmatrix}$ is mathematically equivalent to $\frac{1}{\sqrt{2}}(\alpha+\beta)\frac{1}{\sqrt{2}}\begin{bmatrix}1\\1\end{bmatrix} + \frac{1}{\sqrt{2}}(\alpha-\beta)\frac{1}{\sqrt{2}}\begin{bmatrix}1\\-1\end{bmatrix}$. It just takes a little algebra to verify this equivalence:

$$\frac{1}{\sqrt{2}}(\alpha+\beta)\frac{1}{\sqrt{2}}\begin{bmatrix}1\\1\end{bmatrix} = \frac{1}{2}(\alpha+\beta)\begin{bmatrix}1\\1\end{bmatrix} = \frac{1}{2}\begin{bmatrix}\alpha+\beta\\\alpha+\beta\end{bmatrix}$$

and

$$\frac{1}{\sqrt{2}}(\alpha-\beta)\frac{1}{\sqrt{2}}\begin{bmatrix}1\\-1\end{bmatrix} = \frac{1}{2}(\alpha-\beta)\begin{bmatrix}1\\-1\end{bmatrix} = \frac{1}{2}\begin{bmatrix}\alpha-\beta\\\beta-\alpha\end{bmatrix}.$$

Adding these two together gives $\frac{1}{2}\begin{bmatrix}\alpha+\beta+\alpha-\beta\\ \alpha+\beta+\beta-\alpha\end{bmatrix}=\frac{1}{2}\begin{bmatrix}2\alpha\\ 2\beta\end{bmatrix}=\begin{bmatrix}\alpha\\ \beta\end{bmatrix}$. So just as every spinor can be expressed as a complex weighted sum of a z-spin-up piece and a z-spin-down piece, so it can be expressed as the complex weighted sum of an x-spin-up piece and an x-spin-down piece.

Predicting what will happen if we send an electron represented by a certain spinor through an x-oriented Stern-Gerlach magnet proceeds in exactly the same way as for z-spin: first express the spinor as $\gamma|x\uparrow>+\delta|x\downarrow>$ with $|\gamma|^2+|\delta|^2=1$, then use $|\gamma|^2$ as the probability that the electron will be deflected upward and $|\delta|^2$ as the probability it will be deflected downward. And the same game can be played with spin in the y-direction, using $|y\uparrow>=\frac{1}{\sqrt{2}}\begin{bmatrix}1\\ i\end{bmatrix}$ and $|y\downarrow>=\frac{1}{\sqrt{2}}\begin{bmatrix}1\\ -i\end{bmatrix}$. The same sort of thing can be done for handling Stern-Gerlach magnets oriented in any intermediate direction.

This method for making predictions about experiments with Stern-Gerlach magnets has the Heisenberg uncertainty relations built in. Suppose, for example, we manage to prepare a beam of electrons so that every electron is deflected upward by a z-oriented magnet. The spinor associated with each electron in the beam must be $|z\uparrow>=\begin{bmatrix}1\\ 0\end{bmatrix}$. What if we pass the beam through an x-oriented magnet? To make the prediction, we write $\begin{bmatrix}1\\ 0\end{bmatrix}$ as $\frac{1}{\sqrt{2}}|x\uparrow>+\frac{1}{\sqrt{2}}|x\downarrow>$ (check that this is right!). The recipe then predicts that there is a 50% chance of the electron being deflected up and a 50% chance of it being deflected down. If we are using the predictive recipe and are certain how an electron will be influenced by a z-oriented magnet, we must also be maximally uncertain about how it will be influenced by a magnet oriented in the x-direction (and similarly uncertain about the y-direction). If we are using the predictive recipe, then it is *mathematically impossible* to write down a spinor that allows us to make predictions with certainty in more than one direction, and the more certain we get in one direction the more uncertain we will be in the orthogonal directions. It is easy to see how all the phenomena observed in Experiment 5 are predicted by the recipe.

The astute reader may at this point feel a nagging worry. The predictive recipe has three steps: 1) assign an initial wavefunction to the system, 2) use Schrödinger's equation to evolve that wavefunction in time, and 3) use Born's rule to assign probabilities to the outcome if a position measurement is made. But in describing how to deal with Experiment 5, we have used language like "if the electron is passed through a Stern-Gerlach magnet oriented in the z-direction." It is not immediately clear how this kind of information gets incorporated into the recipe at all.

The only place that the presence of a particular magnet can affect the recipe is in the potential term in Schrödinger's equation. It is this term that reflects how the electron interacts with other things. So a particularly configured Stern-Gerlach magnet (and the magnetic field it produces) makes its influence felt in this potential term. Just for show, the Stern-Gerlach interaction in the Hamiltonian is represented by the term $-\frac{e}{m}(\frac{\hbar}{2}\vec{\sigma}) \cdot \vec{B}$, where e and m are respectively the charge and mass of the electron, and σ represents the spinor and $\vec{B}$ the magnetic field. Changing the direction of magnetic field changes the way this term influences the evolution of the wavefunction.

When we do a spin experiment or measure the z-spin of a particle, the outcome is always an event that happens at one place rather than another: a flash, for example, occurs in one region of the screen rather than another. But predictions for the location of flashes in space are derived, via Born's rule, from the spatial part of the wavefunction rather than directly from its spinor part. This happens exactly because, by means of the sort of potential written above, the spin degrees of freedom can become entangled with the spatial degrees of freedom.

Here's a quick example. We know that if we feed a z-spin up electron through a Stern-Gerlach magnet oriented in the z-direction, the outcome will be that the whole spatial part of the wavefunction will be deflected upward: there is no chance to later find the electron deflected down. Schematically, $|z\uparrow\rangle|\text{middle}\rangle \Rightarrow |z\uparrow\rangle|\text{upward}\rangle$, where $|\text{middle}\rangle$ represents a spatial wavefunction for a beam directed at the middle region of the Stern-Gerlach magnet, $|\text{upward}\rangle$ represents a spatial wavefunction of a beam directed upward, and the arrow $\Rightarrow$ represents time evolution of the wavefunction generated by Schrödinger's equation. Similarly, $|z\downarrow\rangle|\text{middle}\rangle \Rightarrow |z\downarrow\rangle|\text{downward}\rangle$. Each of these wavefunctions is a product state of a spinor part and a spatial part. But what if we feed an x-spin up beam of electrons in?

The linearity of Schrödinger's equation again does the job. The initial x-spin up beam is represented by $|x\uparrow>|\text{middle}>$, which is the same as $(\frac{1}{\sqrt{2}}|z\uparrow> + \frac{1}{\sqrt{2}}|z\downarrow>)|\text{middle}>$, which is the same as $\frac{1}{\sqrt{2}}|z\uparrow>|\text{middle}> + \frac{1}{\sqrt{2}}|z\downarrow>|\text{middle}>$. But we know how Schrödinger's equation evolves each of these pieces separately. So the evolution of the sum is just the sum of the evolutions:

$$\frac{1}{\sqrt{2}}|z\uparrow>|\text{middle}> + \frac{1}{\sqrt{2}}|z\downarrow>|\text{middle}> \Rightarrow$$
$$\frac{1}{\sqrt{2}}|z\uparrow>|\text{upward}> + \frac{1}{\sqrt{2}}|z\downarrow>|\text{downward}>.$$

In this last state, the spin part of the wavefunction has become entangled with the spatial part. Using Born's rule, we predict a 50% chance of a flash occurring in the upper region of the screen and a 50% chance of a flash in the lower region.

These same rules also allow us to predict outcomes of Experiment 6 with the Mach-Zehender interferometer. As long as we do not appeal to Born's rule, the evolution of the wavefunction is governed by Schrödinger's equation, and spatially separated parts of wavefunctions can be recombined to predictable effect. Letting $|\text{upper}>$ now mean "proceeding along the upper path of the interferometer" and $|\text{lower}>$ mean "proceeding along the lower path," our x-spin up beam fed into the interferometer will evolve into $\frac{1}{\sqrt{2}}|z\uparrow>|\text{upper}> + \frac{1}{\sqrt{2}}|z\downarrow>|\text{lower}>$. But the structure of the interferometer allows the two paths to reconverge:

$$|z\uparrow>|\text{upper}> \Rightarrow |z\uparrow>|\text{diagonal}>, \text{ and}$$
$$|z\downarrow>|\text{lower}> \Rightarrow |z\downarrow>|\text{diagonal}>,$$

where $|\text{diagonal}>$ indicates a spatial trajectory along the diagonal path at the top of Figure 10. By linearity (once again), we get

$$\frac{1}{\sqrt{2}}|z\uparrow>|\text{middle}> + \frac{1}{\sqrt{2}}|z\downarrow>|\text{middle}> \Rightarrow$$
$$\frac{1}{\sqrt{2}}|z\uparrow>|\text{upper}> + \frac{1}{\sqrt{2}}|z\downarrow>|\text{lower}> \Rightarrow$$

$$\frac{1}{\sqrt{2}}|z\uparrow>|\text{diagonal}> + \frac{1}{\sqrt{2}}|z\downarrow>|\text{diagonal}> =$$

$$\left(\frac{1}{\sqrt{2}}|z\uparrow> + \frac{1}{\sqrt{2}}|z\downarrow>\right)|\text{diagonal}> = |x\uparrow>|\text{diagonal}>.$$

We recover a beam of pure $|x\uparrow>$ and we *disentangle* the spin from spatial degrees of freedom. No surprise that if we run the recombined beam through an x-oriented magnet, it all gets deflected upward.

What about Albert's magical "total of nothing" box? It too can be straightforwardly treated. The effect of the applied magnetic field is to change the phase of the spinor of a particle that passes through it. In particular, the phase is changed by multiplying by -1: $\left[\begin{smallmatrix}1\\0\end{smallmatrix}\right]$ is converted to $\left[\begin{smallmatrix}-1\\0\end{smallmatrix}\right]$, $\frac{1}{\sqrt{2}}\left[\begin{smallmatrix}1\\1\end{smallmatrix}\right]$ changes to $\frac{1}{\sqrt{2}}\left[\begin{smallmatrix}-1\\-1\end{smallmatrix}\right]$, and so forth. Now if one applies such a change of phase to an entire wavefunction, the predictive recipe will give exactly the same statistical predictions, since Born's rule requires us to take the squared amplitude of the wavefunction and the squaring operation yields exactly the same result for $|\psi|^2$ as for $|-\psi|^2$.

But in our experimental configuration, we do not run the whole beam through the magnetic field: only the part on the lower path goes through. As a result, $\frac{1}{\sqrt{2}}|z\downarrow>|\text{lower}>$ changes into $-\frac{1}{\sqrt{2}}|z\downarrow>|\text{lower}>$. When the two beams recombine, the calculation now yields:

$$\frac{1}{\sqrt{2}}|z\uparrow>|\text{upper}> - \frac{1}{\sqrt{2}}|z\downarrow>|\text{lower}> \Rightarrow$$

$$\frac{1}{\sqrt{2}}|z\uparrow>|\text{diagonal}> - \frac{1}{\sqrt{2}}|z\downarrow>|\text{diagonal}> =$$

$$\left(\frac{1}{\sqrt{2}}|z\uparrow> - \frac{1}{\sqrt{2}}|z\downarrow>\right)|\text{diagonal}> = |x\downarrow>|\text{diagonal}>.$$

The recipe predicts that every electron passing through the interferometer-plus-phase-shifter should be deflected downward by an x-oriented magnet.

Note that to derive these predictions, it is essential that one not employ Born's rule in the following way while the electron is en route through the interferometer. Suppose one thought that when the recipe yields the state $\frac{1}{\sqrt{2}}|z\uparrow>|\text{upper}> - \frac{1}{\sqrt{2}}|z\downarrow>|\text{lower}>$, the actual physical state of the electron must be properly described by either |*z*↑>upper> or |*z*↓>|lower>, with Born's rule supplying a 50% probability of each. Such an application of Born's rule leads to trouble. For if the state is really |*z*↑>|upper>, then the electron should have a 50% chance of being deflected upward by an *x*-oriented magnet if it encounters one later on, and if the state is really |*z*↓>|lower>, then also it has a 50% chance of upward deflection by an *x*-oriented magnet. Either way, the electron should have a 50% chance of upward deflection. But in this experiment, that result is just empirically wrong: 100% of the electrons are deflected downward and none upward.

(It does not follow that there is no definite fact about which path the electron takes through the device! As we will see in chapter 5, according to one precise theory, each electron takes either only the upper path or only the lower path, with about 50% going each way. But according to this theory, the complete physical state of the electron is not described by either |*z*↑>|upper> or |*z*↓>|lower> on any particular run.)

Albert's "total-of-nothing" phase-shifting device illustrates one important aspect of the predictive recipe: if some circumstance results in changes of phase in part of the wavefunction, this can result in empirically observable changes via interference. But what is relevant mathematically must be a change in the relative phases of two parts of the wavefunction when they are recombined. The relative phase determines which parts of the wavefunction interfere constructively rather than destructively. In a simple two-slit water-table experiment, for example, systematically changing the phase of the water coming through one slit (e.g., changing crests into troughs) will move the places where there is constructive and destructive interference at the screen. The interference bands will shift.

Water waves, being described by real numbers, have a sort of absolute phase: There are precise regions where the highest amplitudes (crests) and the lowest amplitudes (troughs) of the waves occur. The phase of the wavefunction, as a complex field, is different: the magnitude does not change, so there is no crest or trough. Mathematically, we ascribe a phase, $e^{i\theta}$, to each point in configuration space, but for predictive purposes, it is only the difference in phases that matter when two parts of the wavefunction are brought together by the Schrödinger evolution.

Eigenstates, Eigenvalues, Hermitian Operators, and All That

The previous section treating spin avoids the technical apparatus that appears in standard introductions to quantum theory. Let us pause here to explore that apparatus. I will also explain why we have been avoiding it.

I have already introduced the notion of an operator on wavefunctions: An operator maps an input wavefunction to an output wavefunction. We represent operators by capital letters with hats, so $\widehat{O}$ can stand for a generic operator. Operators can have various mathematical properties. One important property is *linearity*. If $\widehat{O}$ is linear, then operating on the superposition of two wavefunctions gives the same result as operating on the wavefunctions individually and then superposing the results. That is, for a linear operator, we have

$$\widehat{O}(\alpha\,|\psi> + \beta\,|\phi>) = \alpha\widehat{O}\,|\psi> + \beta\widehat{O}\,|\phi>.$$

Linearity is a very important property to keep track of. The Hamiltonian operator, which generates the time evolution of the wavefunction in the Schrödinger equation, is a linear operator, and that mathematical property lies at the heart of some of the central interpretive problems for quantum mechanics. The famous Schrödinger cat argument relies only on the linearity of the Hamiltonian.

Given any operator $\widehat{O}$, we can ask whether there are wavefunctions with the following property: $\widehat{O}\,|\psi> = \varepsilon\,|\psi>$ for some complex number ε. When this property holds, we say that $|\psi>$ is an *eigenfunction* or *eigenstate* of $\widehat{O}$ and that ε is its *eigenvalue*. Knowing the eigenfunctions of a linear operator can simplify doing calculations. For example, suppose we want to know how a particular wavefunction $|\psi>$ will evolve in time according to the Schrödinger equation. We need to know the effect of operating with the Hamiltonian operator $\widehat{H}$ on $|\psi>$. Let $|\phi_1>$, $|\phi_2>$, $|\phi_3>$, . . . , $|\phi_N>$ be a collection of eigenfunctions of $\widehat{H}$ with eigenvalues ε_1, ε_2, ε_3, . . . , ε_N, respectively. If we can write $|\psi>$ as a superposition of the eigenstates, then the calculation becomes simple:

$$\begin{aligned} \widehat{H}\,|\psi> &= \widehat{H}(\alpha\,|\phi_1> + \beta\,|\phi_2> + \gamma\,|\phi_3> + \cdots + \delta\,|\phi_N>) \\ &= \alpha\widehat{H}\,|\phi_1> + \beta\widehat{H}\,|\phi_2> + \gamma\widehat{H}\,|\phi_3> + \cdots + \delta\widehat{H}\,|\phi_N> \text{ (by linearity)} \\ &= \alpha\varepsilon_1|\phi_1> + \beta\varepsilon_2|\phi_2> + \gamma\varepsilon_3|\phi_3> + \cdots + \delta\varepsilon_N|\phi_N>. \end{aligned}$$

If an operator is also *Hermitian* or *self-adjoint*, then its eigenvalues are guaranteed to be real numbers rather than complex numbers with an imaginary part.

In some approaches to understanding quantum theory, great interpretive weight is put on Hermitian operators. The "observable properties" of systems are to be somehow associated with Hermitian operators. Furthermore, when one "measures" such a property, it is said, the possible outcomes of the measurement correspond to the eigenvalues of the operator. To assign probabilities to the various outcomes, one expresses the wavefunction of the system under consideration as a superposition of eigenstates of the operator. The probability of getting a particular outcome is the square of the amplitude assigned to the corresponding eigenstate.

Here is a concrete example of this approach. The Hamiltonian operator is not just the generator of the time evolution of the wavefunction, it is also the operator usually associated with the classical quantity known as the total energy of the system. So suppose we have a laboratory situation set up to measure the total energy of a system, which happens to be assigned the wavefunction $|\psi>$ above. We are then to conclude that the outcome of this experiment must be one of the numbers $\varepsilon_1, \varepsilon_2, \varepsilon_3, \ldots, \varepsilon_N$, and that the chance of getting ε_1 is $|\alpha|^2$, the chance of getting ε_2 is $|\beta|^2$, and so forth. It follows that if a wavefunction predicts with certainty what the outcome of such an energy measurement will be, then the wavefunction is an eigenstate of the Hamiltonian operator.

This same approach applies to spin. For example, we can associate the z-spin of an electron with the matrix $\left[\begin{smallmatrix}1 & 0\\ 0 & -1\end{smallmatrix}\right]$, the x-spin with the matrix $\left[\begin{smallmatrix}0 & 1\\ 1 & 0\end{smallmatrix}\right]$, and the y-spin with the matrix $\left[\begin{smallmatrix}0 & -i\\ i & 0\end{smallmatrix}\right]$. These are called the *Pauli spin matrices*. Each of these 2×2 matrices is an operator on spinors, where the operation is implemented by matrix multiplication. The operations of the three matrices on an arbitrary spinor are as follows:

$$\begin{bmatrix} 1 & 0 \\ 0 & -1 \end{bmatrix}\begin{bmatrix} a \\ b \end{bmatrix} = \begin{bmatrix} a \\ -b \end{bmatrix}, \begin{bmatrix} 0 & 1 \\ 1 & 0 \end{bmatrix}\begin{bmatrix} a \\ b \end{bmatrix} = \begin{bmatrix} b \\ a \end{bmatrix}, \begin{bmatrix} 0 & -i \\ i & 0 \end{bmatrix}\begin{bmatrix} a \\ b \end{bmatrix} = \begin{bmatrix} -ib \\ ia \end{bmatrix}.$$

It is now easy to verify that normalized eigenstates of the z-spin matrix are $\left[\begin{smallmatrix} 1 \\ 0 \end{smallmatrix}\right]$ and $\left[\begin{smallmatrix} 0 \\ 1 \end{smallmatrix}\right]$ with eigenvalues 1 and −1, respectively. The eigenstates and eigenvalues of the x-spin and y-spin matrices are left as exercises.

The problem with this whole standard approach—and the reason we have ignored it—is that there is no prospect of using it to answer our basic questions. Consider, for example, Experiment 5. As a piece of physics, that experiment essentially involves the precise geometry and orientation of a Stern-Gerlach magnet and the magnetic field it creates. A truly fundamental and universal physics ought to treat this situation via physical description, irrespective of conceptualizing it as the "measurement" of anything. As such, what we want to account for is *how certain marks or flashes are formed in certain places on a screen*. This requires providing a physical characterization of the situation. But the approach outlined above short-circuits all the real gritty physics. Rather, we are invited to just somehow conceptualize the entire physical situation as (for example) a z-spin measurement, and the occurrence of a flash or mark as an outcome, and to assign statistics to the possible outcomes by the calculation outlined above. But none of these conceptual characterizations follows in any rigorous way from the physical description of the laboratory apparatus. In particular, since the various observable outcomes of the experiment differ by the location of various marks or flashes in space, we should demand a story about how the spatial aspects of the wavefunction become entangled with the spin aspects. The standard approach sidesteps all of this by mere stipulation: We are told to regard the physical set-up as a measurement but are not told why this assumption is legitimate, or how to determine whether some other laboratory arrangement is a measurement, and if so, of what.

So let us strongly reject the treatment of Hermitian operators, eigenstates, eigenvalues, and so forth as not-further-analyzable mathematical representations of concrete laboratory situations. The laboratory is a physical entity, and it should be subject to physical analysis. It must be that the eventual upshot of the physical analysis accounts for the calculational utility of Hermitian operators, eigenstates, eigenvalues, and the like. If we simply identify (without any further justification) a magnet with a certain geometry and orientation together with a phosphorescent screen as a z-spin measuring device, and associate the whole contraption with the matrix $\begin{bmatrix} 1 & 0 \\ 0 & -1 \end{bmatrix}$, and further identify a flash in one region of the screen as a z-spin up outcome and a flash in another region as a z-spin down outcome, then the standard quantum recipe can be used to make predictions for an electron associated with a given wavefunction. But a completed physics should illuminate why just this sort of physical situation ought to be treated with this particular mathematics. The standard approach systematically hides this basic physical question from view.

We have been trying to stay true to the idea that physics is the theory of matter in motion (i.e., the theory that treats the disposition of matter in space-time). Following this approach, the outcomes of the experiments must be determined by where some matter ends up. For the physics to account for different outcomes, then, it must provide predictions for the locations of things. In the case of our spin experiments, unlike the first four experiments, this required coupling the spinorial part of the wavefunction to the spatial part and then applying Born's rule to the spatial part. The Hamiltonian associated above with the Stern-Gerlach apparatus illustrates how this can be done.

Multiple Particles with Spin

We now have all the pieces in place to apply the recipe to Bohm's version of the EPR experiment and to derive predictions of violations of Bell's inequality. The main work is done by an entangled spin state of two or three electrons. As usual, we construct the entangled state by starting with unentangled product states.

Suppose we have a pair of electrons that begin in the same location, with one traveling off to the right the other to the left. The one going to the right can have the spinor $|z\uparrow>$ and the one traveling to the left $|z\downarrow>$. The resulting product state could be symbolized as

$|z\uparrow, \text{right}>|z\downarrow, \text{left}>$, indicating both the spatial and spin features of the wavefunction for each electron. Similarly, the spin states could be switched: $|z\downarrow, \text{right}>|z\uparrow, \text{left}>$. There is no entanglement in either of these states, and making predictions from them is easy. For example, in the first state, if both particles are passed through z-oriented magnets, the right-moving particle will be deflected up and the left-moving one down. If they are both passed through x-oriented magnets, then each has a 50-50 chance of being deflected either way, with no correlations predicted between them. That is, finding out which direction one goes will not change the prediction about the other. It will still be 50-50.

By the superposition principle, we can form from this pair of states the entangled state

$$\frac{1}{\sqrt{2}}|z\uparrow, \text{right}>|z\downarrow, \text{left}> - \frac{1}{\sqrt{2}}|z\downarrow, \text{right}>|z\uparrow, \text{left}>.$$

This is called the *singlet state* of spin. For convenience, let us indicate the spatial part of the wavefunction by just a subscript and write the singlet state as

$$\frac{1}{\sqrt{2}}|\text{z}\uparrow>_{\text{R}}|\text{z}\downarrow>_{\text{L}} - \frac{1}{\sqrt{2}}|\text{z}\downarrow>_{\text{R}}|\text{z}\uparrow>_{\text{L}}.$$

What should we predict if we pass both electrons through z-oriented magnets followed by a phosphorescent screen?

The magnets will entangle the spinor of each electron with its spatial wavefunction. Recall that the spatial part of the wavefunction is defined over the configuration space of the system. In configuration space, after the electrons pass the magnets, there will be a lump of the wavefunction in the region corresponding to the right-hand particle being deflected up and the left-hand particle down, and an equal-amplitude lump corresponding to the right being deflected down and the left up. So by Born's rule, we predict a 50% chance of the right-hand flash occurring up and the left down, and a 50% chance of the right-hand flash being down and the left up. There is no chance that both with will be up or both down. In short, is it certain that the location of one flash will be up and the other down, but completely uncertain which will be up and which down. Observing either flash renders one completely sure of where the other will be.

Einstein argued that in this case, where the two electrons can be arbitrarily far apart from each other, we cannot accept that what happens to one electron can have any physical influence or effect on the other. But absent such "spooky action-at-a-distance," it follows that each electron must be predisposed all along to be deflected the way it is: otherwise, how could the second electron, uninfluenced by the first, always behave the opposite way? The predictive recipe does not specify which electron will go which way, so Einstein's conclusion is that the predictive recipe must not be representing all the physical facts. The wavefunction evidently (he argued) does not actually reflect all the physical characteristics of the electrons. A completed physics should do better.

But even odder things happen. What if, instead of orienting both magnets in the z-direction, we orient them in the x-direction? We have the resources to answer this: just rewrite the singlet state in terms of x-spin rather than z-spin:

$$\frac{1}{\sqrt{2}}|z\uparrow>_R|z\downarrow>_L - \frac{1}{\sqrt{2}}|z\downarrow>_R|z\uparrow>_L =$$

$$\frac{1}{\sqrt{2}}\left(\frac{1}{\sqrt{2}}|x\uparrow>_R + \frac{1}{\sqrt{2}}|x\downarrow>_R\right)\left(\frac{1}{\sqrt{2}}|x\uparrow>_L - \frac{1}{\sqrt{2}}|x\downarrow>_L\right)$$

$$-\frac{1}{\sqrt{2}}\left(\frac{1}{\sqrt{2}}|x\uparrow>_R - \frac{1}{\sqrt{2}}|x\downarrow>_R\right)\left(\frac{1}{\sqrt{2}}|x\uparrow>_L + \frac{1}{\sqrt{2}}|x\downarrow>_L\right) =$$

$$\frac{1}{2\sqrt{2}}(|x\uparrow>_R|x\uparrow>_L - |x\uparrow>_R|x\downarrow>_L + |x\downarrow>_R|x\uparrow>_L - |x\downarrow>_R|x\downarrow>_L) -$$

$$\frac{1}{2\sqrt{2}}(|x\uparrow>_R|x\uparrow>_L + |x\uparrow>_R|x\downarrow>_L - |x\downarrow>_R|x\uparrow>_L - |x\downarrow>_R|x\downarrow>_L) =$$

$$\frac{1}{2\sqrt{2}}(-2|x\uparrow>_R|x\downarrow>_L + 2|x\downarrow>_R|x\uparrow>_L) =$$

$$\frac{-1}{\sqrt{2}}(|x\uparrow>_R|x\downarrow>_L - |x\downarrow>_R|x\uparrow>_L).^{6}$$

[6]If you, dear reader, are anything like the author, your eyes have just glazed over and you have decided to give me credit for getting my math right and skipped the details. Please don't! It's just a little painless algebra that anyone can do, and

Except for the factor of –1 (which makes no difference to the Born's rule predictions, as we have seen), the singlet state has exactly the same mathematical form when expressed in terms of x-spin as it does when expressed in terms of z-spin. So the predictions for the case of two x-oriented magnets are just the same: 50% that the right electron is deflected up and the left down, and 50% that the right is deflected down and the left up. And if Einstein's argument works for z-spin, it also works for x-spin: absent spooky-action-at-a-distance, the way each electron would react to an x-oriented magnet must be physically predetermined and independent of what happens to the other electron. But now the quantum formalism is in serious trouble: Each electron would have to have a predetermined z-spin and a predetermined x-spin, but no spinor permits simultaneous prediction of both with certainty. Therefore, Einstein would conclude, the wavefunction must be leaving something out.

Not only can we get the predicted perfect correlations of the EPR argument out of the predictive recipe, we can also make predictions when the magnets on the two sides are misaligned. This sort of set-up was originally used by Bell to prove his result.

There are two ways to approach this. Suppose the magnet on the left is oriented in the z-direction and the one on the right is oriented in the z-x plane but is offset by 60° from the z-direction. It would then be most convenient to rewrite the singlet state in yet another way, in terms of z-spin on the left and 60°-spin on the right. The 60°-spin-up spinor is $\begin{bmatrix}\sqrt{3}/2\\1/2\end{bmatrix}$, and the spin-down spinor is $\begin{bmatrix}1/2\\-\sqrt{3}/2\end{bmatrix}$. (In general, the spin-up spinor for a magnet oriented at an angle θ in the z-x plane is $\begin{bmatrix}\cos(\theta/2)\\\sin(\theta/2)\end{bmatrix}$, and the spin-down spinor is $\begin{bmatrix}\sin(\theta/2)\\-\cos(\theta/2)\end{bmatrix}$. If the magnet is not oriented in the z-x plane, then the spinor will contain some imaginary components.) So we have $|60°\uparrow> = \frac{\sqrt{3}}{2}|z\uparrow> + \frac{1}{2}|z\downarrow>$ and $|60°\downarrow> = \frac{1}{2}|z\uparrow> - \frac{\sqrt{3}}{2}|z\downarrow>$. Solving for $|z\uparrow>$ and $|z\downarrow>$ in terms of $|60°\uparrow>$ and $|60°\downarrow>$ yields

---

doing it produces a sense of both accomplishment and understanding that can be acquired in no other way.

$$|z\uparrow\rangle = \frac{\sqrt{3}}{2}|60°\uparrow\rangle + \frac{1}{2}|60°\downarrow\rangle \text{ and}$$

$$|z\downarrow\rangle = \frac{1}{2}|60°\uparrow\rangle - \frac{\sqrt{3}}{2}|60°\downarrow\rangle.$$

The singlet state $\frac{1}{\sqrt{2}}|z\uparrow\rangle_R|z\downarrow\rangle_L - \frac{1}{\sqrt{2}}|z\downarrow\rangle_R|z\uparrow\rangle_L$ is therefore mathematically the same as

$$\frac{1}{\sqrt{2}}\left(\frac{\sqrt{3}}{2}|60°\uparrow\rangle_R + \frac{1}{2}|60°\downarrow\rangle_R\right)|z\downarrow\rangle_L -$$
$$\frac{1}{\sqrt{2}}\left(\frac{1}{2}|60°\uparrow\rangle_R - \frac{\sqrt{3}}{2}|60°\downarrow\rangle_R\right)|z\uparrow\rangle_L =$$
$$\frac{\sqrt{3}}{\sqrt{8}}|60°\uparrow\rangle_R|z\downarrow\rangle_L + \frac{1}{\sqrt{8}}|60°\downarrow\rangle_R|z\downarrow\rangle_L -$$
$$\frac{1}{\sqrt{8}}|60°\uparrow\rangle_R|z\uparrow\rangle_L + \frac{\sqrt{3}}{\sqrt{8}}|60°\downarrow\rangle_R|z\uparrow\rangle_L.$$

The empirical predictions for our experimental situation—with the magnet on the left set in the z-direction and the magnet on the right in the 60°-direction—can be read off this state. The magnets will entangle the spin degrees of freedom with the spatial degrees so that, for example, $|60°\uparrow\rangle_R$ gets associated with the spatial part of the wavefunction for the right-hand electron propagating along the "up" output channel. By Born's rule, the amplitudes $\frac{\sqrt{3}}{\sqrt{8}}$, $\frac{1}{\sqrt{8}}$, and $-\frac{1}{\sqrt{8}}$ are squared to yield the probabilities for the four possible outcomes: $\frac{3}{8}$ chance that the flash on the right occurs in the "up" region of the 60° apparatus and the flash on the left occurs in its "down" region; $\frac{1}{8}$ that both occur down; $\frac{1}{8}$ that both occur up; and $\frac{3}{8}$ that the flash on the right occurs in the "down" region and the flash on the left occurs in the "up" region. Overall, there is a $\frac{3}{4}$ chance that the flashes give opposite results and $\frac{1}{4}$ that they give the same result.

What if we decide to orient the magnet on the left in the z-direction but have not yet decided what, if anything, to do on the right? Just looking at the singlet state $\frac{1}{\sqrt{2}}|z\uparrow\rangle_R|z\downarrow\rangle_L - \frac{1}{\sqrt{2}}|z\downarrow\rangle_R|z\uparrow\rangle_L$, we would naturally take Born's rule to yield a 50% chance of each possible outcome on the left. Now suppose that, in fact, the flash on the left is "up." How do we take account of that result in the recipe to make further predictions about the right-hand electron?

The standard mathematical procedure is called *collapse of the wavefunction*. That is, given that the initial wavefunction is $\frac{1}{\sqrt{2}}|z\uparrow>_R|z\downarrow>_L - \frac{1}{\sqrt{2}}|z\downarrow>_R|z\uparrow>_L$ and given that the flash on the left was up, we simply discard the term of the wavefunction containing $|z\downarrow>_L$. We are then left with just $-\frac{1}{\sqrt{2}}|z\downarrow>_R|z\uparrow>_L$, which gets renormalized to $|z\downarrow>_R|z\uparrow>_L$. Our entangled state has become a product state, and we conclude that, for further predictive purposes, the right-hand electron can be ascribed the state $|z\downarrow>_R$. It is easy to check that this procedure yields the following conditional prediction: If the right-hand magnet is oriented in the z-direction and the flash is in the "up" region, then the probability for a the left-hand flash to be down should that electron be passed through a z-oriented magnet is 1, and the probability for the flash to be down if passed through a 60°-oriented magnet is $\frac{3}{4}$.

This collapse of the wavefunction cannot be produced by Schrödinger evolution of the two-particle system—the linearity of the Schrödinger equation ensures that. Indeed, the exact physical implications of this mathematical procedure are hotly disputed. In some theories, as we will see, the mathematical collapse closely mirrors a real physical evolution. So-called Quantum Bayesians liken it instead to a mere change of beliefs: updating one's subjective degrees of credence on receipt of new information. In other theories, it has quite a different status. But once again, we are not yet in the business of drawing physical or ontological or dynamical conclusions at all: We are just concocting a practical recipe for making predictions.

But the reader might well wonder, doesn't the exact status of this "collapse postulate" have further empirical consequences? After all, we are already acutely aware of how different terms of a wavefunction can (through Schrödinger evolution) come to interfere with each other and thereby produce observable effects. If a term in a wavefunction is just thrown away or annihilated, then, obviously, no further interference can be produced by it. So if the wavefunction always evolves by Schrödinger's equation, and no term ever disappears, surely there must be circumstances in which one gets different predictions than one would get using a mathematical collapse at some point.

This conclusion is correct. In principle, a no-collapse recipe makes different empirical predictions than a collapse recipe. That invites two questions: 1) Does the "official" quantum recipe contain collapse or not? and 2) Why haven't experiments been done to settle whether, to get the right predictions, there needs to be a collapse? The answer to the first question is that the quantum recipe is somewhat vague at exactly this point. Some textbooks directly postulate the collapse (e.g., Shankar 1980, p. 120, postulate III). Others implicitly rely on it, but its status is unclear. In the famous textbook by Landau and Lifshitz (1965), for example, a curiously incoherent story is told in which a "classical apparatus" interacts with a "quantum object." Both are ascribed wavefunctions to which the Schrödinger evolution is applied, but in addition, it is demanded that the "classical apparatus" have a "physical quantity" called the "reading of the apparatus," which must always be definite. By this unprincipled dodge, Landau and Lifshitz are able to "derive" the collapse of the quantum object's wavefunction (see Landau and Lifshitz 1965, pp. 21–22). As to the second question, although in principle there are empirical differences between the predictions of a collapse and a no-collapse recipe, as a practical matter it can be extremely difficult to realize the experimental conditions in which these differences can be checked. Recall that the issue here is how to make further predictions after an application of Born's rule (i.e., after a position measurement has been made). It is not entirely clear which physical situations this refers to, but in practice, such situations involve the use of macroscopic (and hence very complex and complicated) bodies. As a practical matter, manipulating and modeling such bodies with the degree of precision required to manifest these different predictions is an extremely difficult task. So for all practical purposes, it often makes no predictive difference whether one collapses the wavefunction before the final application of Born's Rule. As we have already seen in the Double Slit with Monitoring experiment, entanglement and decoherence tend to destroy interference effects. Collapsing the wavefunction before that final outcome would destroy interference between parts of the superposition as well, but if the linear Schrödinger evolution already has decohered the wavefunction, this is not noticeable.

The only experiment left to discuss is the GHZ experiment. Since it involves three entangled particles, it is a bit more complex than the EPR experiment, and I won't go into all the details. But at least a glance will help.

We know that the GHZ state is an entangled state, and that it predicts (among other things) that if all three magnets are aligned in the x-direction, there will certainly be an odd number of "up" results. Calling the three electrons A, B, and C, the only product spin states that yield this behavior are $|x\uparrow>_A|x\uparrow>_B|x\uparrow>_C$, $|x\uparrow>_A|x\downarrow>_B|x\downarrow>_C$, $|x\downarrow>_A|x\uparrow>_B|x\downarrow>_C$, and $|x\downarrow>_A|x\downarrow>_B|x\uparrow>_C$. The GHZ state must therefore be a superposition of these four states. In fact, an equal superposition of all four with the right phase relations (reflected in the minus signs) serves our purpose:

$$\Psi_{GHZ} = \frac{1}{2}|x\uparrow>_A|x\uparrow>_B|x\uparrow>_C - \frac{1}{2}|x\uparrow>_A|x\downarrow>_B|x\downarrow>_C - \frac{1}{2}|x\downarrow>_A|x\uparrow>_B|x\downarrow>_C - \frac{1}{2}|x\downarrow>_A|x\downarrow>_B|x\uparrow>_C.$$

Each amplitude is either $\frac{1}{2}$ or $-\frac{1}{2}$, so when we square the amplitude to get probabilities, we find that each outcome has a chance of .25. To check what happens if the magnets are aligned in the z-direction for particles A and B but the x-direction for particle C, just replace $|x\uparrow>_A$ and $|x\downarrow>_B$ with their expressions in terms of $|z\uparrow>$ and $|z\downarrow>$ and calculate (see Problem 2 at the end of the chapter).

We have now exposited enough of the quantum recipe to allow us to derive predictions for many experimental situations. Those predictions will be perfectly accurate. If all one wants out of a physical theory is such an accurate prediction-making machine, then there is little need to read the rest of this book. The recipe is a little vague here and there—how to arrive at an initial wavefunction, how to specify a Hamiltonian, and (quite strikingly) when to appeal to Born's Rule—but in practice the vagueness doesn't matter. Appeals to classical physics and "common sense" (e.g., that the flashes on the screen are the sorts of thing that ought to be predicted by Born's Rule) suffice in most circumstances. As John Bell insisted, "ORDINARY QUANTUM MECHANICS (as far as I know) IS JUST FINE FOR ALL PRACTICAL PURPOSES."[7]

What the quantum recipe does not resolve, what it does not even purport to address, is *what the physical world is like such that the quantum recipe works so well.* To answer this question, we need not more recipes, or better recipes, but something quite different from a recipe. We need a physical theory, a clear specification of what there is in the physical world and how it behaves. It is a plain fact about the world that the quantum recipe is an excellent predictive apparatus. That fact needs to be explained. And the recipe itself does not have the right form to serve as an explanation, because it is not a theory. The recipe itself does not say, for example, which parts of the mathematics used in the recipe represent physical features of the world and which do not.

In the next chapter, we begin our examination of several ways to construct a physical theory that explains the success of the quantum recipe. We start with the status of the wavefunction.

Problems

1) Find normalized eigenstates and their eigenvalues for the x-spin and y-spin matrices $\left[\begin{smallmatrix} 0 & 1 \\ 1 & 0 \end{smallmatrix}\right]$ and $\left[\begin{smallmatrix} 0 & -i \\ i & 0 \end{smallmatrix}\right]$.
2) By rewriting the GHZ state, show that if one orients the magnets for particles A and B in the z-direction, the recipe predicts with certainty that there will be an even number of "up" outcomes. Would this still be the case if the original state were $\frac{1}{2}|x\uparrow>_A|x\uparrow>_B|x\uparrow>_C + \frac{1}{2}|x\uparrow>_A|x\downarrow>_B|x\downarrow>_C + \frac{1}{2}|x\downarrow>_A|x\uparrow>_B|x\downarrow>_C + \frac{1}{2}|x\downarrow>_A|x\downarrow>_B|x\uparrow>_C$?
3) Figure 17 depicts an experimental arrangement. A z-oriented Stern-Gerlach magnet is followed by two

[7]Bell (2004), p. 214. Bell employed the acronym FAPP to represent "For All Practical Purposes."

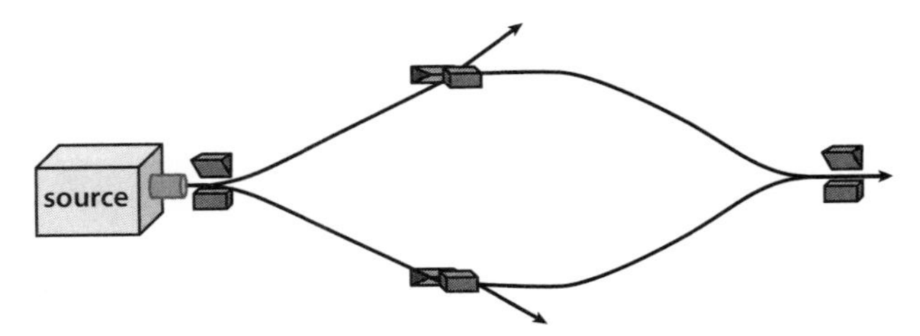

Figure 17

x-oriented magnets, one in each output channel, and then the downward channel of one of these is recombined with the upward channel of the other. The recombined middle beam is then fed through a second z-oriented magnet. Suppose an $|x\uparrow>$ electron is fed into this device. What is the chance that a flash appears in the upward output channel of the second z-oriented magnet? What if an $|x\downarrow>$ electron is fed in? What would happen if the top and bottom output channels were recombined instead of the middle two?

Further Reading

For those readers with some background in mathematics and physics, more particular experimental arrangements are discussed and many more problems offered in Norsen (2017). That book is designed to be an introduction to quantum mechanics with a specific emphasis on foundational and conceptual issues, pitched at the introductory undergraduate level.

CHAPTER 3

The Wavefunction and the Quantum State

THE QUANTUM RECIPE requires that one ascribe a wavefunction[1] to a system, but that fact alone tells us nothing about the physical state or physical properties of the system. Does the wavefunction represent any physical feature of an individual system at all? If so, does it somehow represent all physical features of the system or only some of them? This is the first question to face in trying to draw ontological consequences from the quantum formalism.

A theory according to which the wavefunction does represent some physical feature of an individual system has been called a *ψ-ontic* theory.[2] If the wavefunction does not represent any physical feature of an individual system, what does it represent? Historically, two prominent approaches have sought to deny any sort of direct ontic significance to the wavefunction.

The first, which we can call *ψ-statistical*, maintains that wavefunctions do not properly characterize individual systems at all; instead they only characterize collections of systems. Some sort of "preparation procedure" can be used repeatedly to produce such a collection, and the wavefunction represents characteristics of the collection, not of any particular member of the collection. On such a view, one and the same individual system might correctly be ascribed different wavefunctions depending on which collective of systems it happens to be grouped with. There would be no fact about which wavefunction "really" characterizes the individual.

[1] Some presentations of the quantum recipe allow one to ascribe a more generic mathematical object called a *density matrix* to a system. We will address this wrinkle anon.

[2] This terminology was introduced in Harrigan and Spekkens (2010).

The second non-ontic approach, which we can call *ψ-credal*, maintains that the wavefunction doesn't characterize the physical state of either a single system or a collection of systems. Instead, it characterizes only some cognitive agent's information or beliefs about a system. This approach has been strongly advocated by some researchers in quantum information theory.[3] According to this approach, changes in the wavefunction need not correspond to any physical change in a system at all but only to changes in what someone believes about the system. Sometimes ψ-statistical and ψ-credal are lumped together under the general rubric *ψ-epistemic* theories, although the two approaches are somewhat different.

So at first pass, we have three views: 1) The wavefunction represents a real physical characteristic of an individual system (ψ-ontic), 2) the wavefunction represents only the collective features of an ensemble of systems (ψ-statistical), and 3) the wavefunction represents only some agent's information about an individual system (ψ-credal). In general, a ψ-ontic theory will postulate a wavefunction for the entire universe, representing a real physical feature that exists independently of any cognitive agents, a ψ-statistical theory can make little sense of a wavefunction of the universe (since there is only one) but would be happy to associate wavefunctions with collections of systems shortly after the Big Bang, and a ψ-credal theory can make no sense of wavefunctions ascribed to either individual systems or collections of systems without there also being some agent with beliefs about the systems. In contemporary discussions, both ψ-statistical and ψ-credal would be grouped together into the category of ψ-epistemic theories.

Since any ψ-ontic theory takes the wavefunction to be a mathematical representation of some real physical feature of individual systems, we can introduce a term for that feature: we will call it the *quantum state* of the system.[4] The ψ-statistical and ψ-credal

[3] For an introduction to this line of thought, see Fuchs (2010).

[4] This deliberate use of terminology is not standard. In many texts, the term "quantum state" is used for a mathematical item, an abstract representative of

theories deny that there is any such quantum state of an individual system to be described.

One advocate of a ψ-statistical approach was Einstein. He appreciated the tremendous empirical success of the quantum recipe but, as we have seen, used the EPR set-up together with a commitment to locality (no spooky action-at-a-distance) to argue that the quantum-mechanical description of a system cannot be complete. This conclusion alone does not rule out a ψ-ontic theory: The wavefunction might mathematically represent some real physical feature of an individual system but not every real physical feature. Nonetheless, Einstein was inclined against ψ-ontic approaches and in favor of the idea that the wavefunction characterizes only the statistical properties of collectives. This likely also flowed from a commitment to a different sort of locality that Einstein believed in, called *separability*. Separability requires that the physical states of spatially separated systems be specifiable independently of one another. Entangled wavefunctions, as we have seen, fail to have this feature, while product states do. So taking an entangled wavefunction between distant individual systems as reflecting something physically real would violate Einstein's commitments.

The desire to deny spooky action-at-a-distance also has motivated some ψ-credal approaches. In particular, the collapse of the wavefunction—the sudden global change in the form of the wavefunction as the result of an experimental observation—hardly seems very worrying if the wavefunction merely represents someone's beliefs or information about a system rather than somehow representing the physical state of the system itself. To refer again to Bertlmann's socks, if I know that Bertlmann always wears different colored socks and then see that his left sock is pink, my beliefs about his right sock suddenly change: I now know that it is not pink. But that change is no change at all in the

---

what I am calling the quantum state. For our purposes, it suffices to remember that a wavefunction is a mathematical item—as "function" testifies—and the quantum state is whatever real physical feature of an individual system (if any) obtains iff the system is represented by a given wavefunction.

physical state of the right sock! It was not-pink all along. I just didn't know it.

Violations of Bell's inequality, such as in the GHZ experiment, demonstrate that things can't be so simple: The GHZ particles can't be imbued with their dispositions to react to different local experimental arrangements all along, with the observer simply being ignorant about what these local dispositions are. And leaving aside issues of locality and Bell's inequality altogether, one might also wonder how any theory that is not ψ-ontic could possibly account for interference phenomena, such as the Double-Slit experiment (Experiment 3). That experiment demonstrates that on every individual run, for every individual particle, something is physically sensitive to the state of both slits (i.e., whether each is open or closed). Because when both are open, there are regions of the screen where no flash can occur, even though one could occur there with only one slit open.

The surprising and somewhat unsettling thing about the wavefunction is how spread out it becomes. Mathematically, the wavefunction has nonzero values over large areas of the screen and along both paths through the interferometer. It would therefore come as a great relief to believe that the wavefunction only represents a collection of electrons that can "spread out" as each of the electrons goes its own separate way, or that the wavefunction only represents our information about where the electron is, which can become spread out as we lose track of where it might have gone. But neither such a spreading of a collection nor such a spreading of our ignorance can account for the interference effects. Interference depends on the various parts of the wavefunction representing parts of something that really exists, which cancel one another out in each individual experiment. The cancellation exists irrespective of what other experiments might be done at other times and places and irrespective of what anyone knows or believes about the experiment. Interference is a real, observable, physical effect, and it requires a real physical cause. Such a cause must behave in a way that corresponds to how the wavefunction behaves. And any such physical item, whose behavior is reflected in that of the wavefunction, is just what we mean by a quantum state.

Similarly, the elimination of the interference bands in Experiment 4 is a real change in the sort of physical behavior displayed by individual systems. Regions of the screen that never display flashes with the proton absent sometimes do with the proton present. The decisive circumstance is just whether the proton is there, not whether anyone knows or believes it is there. The presence of the phase-shifting magnets in the interferometer experiments alters both the wavefunction and the observable behavior of every single electron, again regardless of whether anyone knows about the presence of the magnet. It is hard to see how any ψ-credal approach can account for this phenomenon. And the difference between having the magnetic field present and having it absent does not merely change the collective statistics: Some individual outcomes always occur when it is there that never occur when it is absent. All the phenomena we have considered point forcefully to a ψ-ontic theory.

Fortunately, a theorem by Matthew Pusey, Jonathan Barrett, and Terry Rudolph (2012) has made the case for ψ-ontic theories even more airtight.

The PBR Theorem

The logic of the Pusey, Barrett, and Rudolph's (PBR) theorem is beautifully simple. Let's take the example of a beam of electrons prepared to be z-spin up and a beam prepared to be x-spin up. The question is this: Is there something physically different about every single electron in the first beam compared with every single electron in the second? According to a ψ-statistical approach, this need not be the case. If the wavefunction represents only the overall statistical characteristics of each collection of electrons, then there could be individual electrons in the z-spin-up beam that are in exactly the same physical state as individual electrons in the x-spin-up beam. And then the very same electron could be attributed the state z-spin up when considered as part of one collective but also attributed the state x-spin up when considered as part of a different collective. Indeed, if this never happens—if each individual electron can only be attributed one specific wavefunction—then nothing is left of the ψ-statistical approach. And similarly, nothing is left of any ψ-credal approach either: Each electron would have some unique associated wavefunction no matter what anyone knows or believes about it.

So the target question is as follows. Take two individual systems prepared so that they are attributed different wavefunctions. Could it nonetheless be the case that the two systems are in physically identical states? In our example, could an electron prepared to be z-spin up possibly be in the same physical state as one prepared to be x-spin up? This is not to claim that every electron prepared in the first way would be in the same physical state as every electron prepared in the second, but just that sometimes an electron prepared in the first way is in the same physical state as an electron prepared in the second. Let's call this hypothetical common physical state S.

The argument then proceeds as follows. If electrons produced by the z-spin-up preparation procedure are sometimes in S and electrons produced by the x-spin-up preparation procedure are sometimes in S, then it must occasionally be the case that a pair of electrons each prepared to be z-spin up are in reality both in the state S, and that a pair of electrons each prepared to be x-spin up are in reality both in the state S, and similarly if one is prepared z-spin up and the other x-spin up. In each of these cases, the pair of electrons ends up in the same physical state—both S—even though the wavefunctions ascribed to the pair will be different: $|z\uparrow>_A|z\uparrow>_B$, $|x\uparrow>_A|x\uparrow>_B$, $|z\uparrow>_A|x\uparrow>_B$, and $|x\uparrow>_A|z\uparrow>_B$. If in reality, the electrons are both in S in each case, the outcome of any further experiment cannot depend on which preparation was used. If an electron is really in state S after a preparation, then its subsequent behavior depends only on the fact that it is in S. How it happened to get that way becomes physically irrelevant.

Now comes the key observation. If a pair of electrons both in state S might have been prepared as $|z\uparrow>_A|z\uparrow>_B$, or as $|x\uparrow>_A|x\uparrow>_B$, or as $|z\uparrow>_A|x\uparrow>_B$, or as $|x\uparrow>_A|z\uparrow>_B$, and if the quantum recipe always makes good predictions, then the subsequent behavior of the pair must be consistent with each of the four possible preparations.

In particular, the pair cannot yield any outcome that is given zero probability by any of the four possible preparations. For if it did, and the pair happened in fact to be prepared in the "forbidden" way, then the quantum recipe would make a bad prediction.

The final step of the argument shows that there exists an experimental procedure on the pair of particles that, according to the quantum recipe, has exactly four possible outcomes and *each outcome is inconsistent with one of the four possible prepared wavefunctions*. That is, one outcome should (according to the recipe) be impossible if the pair has the wavefunction $|z\uparrow>_A|z\uparrow>_B$, another impossible if it has the wavefunction $|x\uparrow>_A|x\uparrow>_B$, and so on. So in this circumstance, the pair of particles can't behave in a way consistent with all four possible preparations. Whatever the system does, that behavior is ruled out by the quantum recipe, given one of the four possible preparations. Ergo it can't be that sometimes an electron prepared to be z-spin up is in exactly the same state as one prepared to be x-spin up.

What is really important about the PBR argument is its logic, which we have sketched out. But it happens that we have developed enough of the quantum formalism to see exactly how the argument works in detail. It is good for your soul to follow at least a few of the actual calculations.

We already know what the four possible wavefunctions assigned to our pair of particles might be. What sort of experiment, with four possible outcomes, are we then to carry out? The essential point is that each of the four possible outcomes has an associated eigenstate—a wavefunction guaranteed to give that outcome—but the eigenstates are all entangled states of the pair. These four entangled states are:

$$\alpha: \frac{1}{\sqrt{2}}|z\uparrow>_A|z\downarrow>_B + \frac{1}{\sqrt{2}}|z\downarrow>_A|z\uparrow>_B$$

$$\beta: \frac{1}{\sqrt{2}}|z\uparrow>_A|x\downarrow>_B + \frac{1}{\sqrt{2}}|z\downarrow>_A|x\uparrow>_B$$

$$\gamma: \frac{1}{\sqrt{2}}|x\uparrow>_A|z\downarrow>_B + \frac{1}{\sqrt{2}}|x\downarrow>_A|z\uparrow>_B$$

$$\delta: \frac{1}{\sqrt{2}}|x\uparrow>_A|x\downarrow>_B + \frac{1}{\sqrt{2}}|x\downarrow>_A|x\uparrow>_B.$$

Each of these entangled states is associated with one of the four possible outcomes.

What we now want to show is that each of these outcomes is in turn inconsistent with one of the four possible preparations. To do this, we need to use the quantum recipe to predict, for each preparation, the probability for each of the four outcomes. So we need to express the preparation as a superposition of the four eigenstates listed above. Having done that, we can just read the probabilities for the outcomes off the state: The probability for each outcome is just the squared amplitude of the associated eigenstate.

Let's do this for the possible preparation $|z\uparrow>_A|z\uparrow>_B$. Since this wavefunction is written in terms of z-spin, it is most convenient to rewrite all four entangled states in terms of z-spin as well. Just plug in the appropriate expression for x-spin in terms of z-spin and do some algebra:

$$\beta = \frac{1}{\sqrt{2}}|z\uparrow>_A|x\downarrow>_B + \frac{1}{\sqrt{2}}|z\downarrow>_A|x\uparrow>_B =$$

$$\frac{1}{\sqrt{2}}|z\uparrow>_A\left(\frac{1}{\sqrt{2}}|z\uparrow>_B - \frac{1}{\sqrt{2}}|z\downarrow>_B\right) + \frac{1}{\sqrt{2}}|z\downarrow>_A\left(\frac{1}{\sqrt{2}}|z\uparrow>_B + \frac{1}{\sqrt{2}}|z\downarrow>_B\right) =$$

$$\frac{1}{2}|z\uparrow>_A|z\uparrow>_B - \frac{1}{2}|z\uparrow>_A|z\downarrow>_B + \frac{1}{2}|z\downarrow>_A|z\uparrow>_B + \frac{1}{2}|z\downarrow>_A|z\downarrow>_B.$$

Similar manipulation yields

$$\gamma = \frac{1}{2}|z\uparrow>_A|z\uparrow>_B + \frac{1}{2}|z\uparrow>_A|z\downarrow>_B - \frac{1}{2}|z\downarrow>_A|z\uparrow>_B + \frac{1}{2}|z\downarrow>_A|z\downarrow>_B \text{ and}$$

$$\delta = \frac{1}{\sqrt{2}}|z\uparrow>_A|z\uparrow>_B - \frac{1}{\sqrt{2}}|z\downarrow>_A|z\downarrow>_B.$$

Finally, we can now express $|z\uparrow>_A|z\uparrow>_B$ as a superposition of α, β, γ, and δ:

$$|z\uparrow>_A|z\uparrow>_B = \frac{1}{2}\beta + \frac{1}{2}\gamma + \frac{1}{\sqrt{2}}\delta.$$

The quantum recipe therefore predicts (square the amplitude!) that if we do the experiment on a pair prepared with the wavefunction $|z\uparrow>_A|z\uparrow>_B$, there is a .25 chance of getting the outcome associated with state β, a .25 chance of the outcome associated with state γ, a .5 chance of getting the outcome associated with state δ, and no chance at all of getting the outcome associated with state α.

Similar calculations reveal that $|x\uparrow>_A|x\uparrow>_B$ cannot yield the outcome associated with δ, $|z\uparrow>_A|x\uparrow>_B$ cannot yield the outcome associated with β, and $|x\uparrow>_A|z\uparrow>_B$ cannot yield the outcome associated with γ. But now the idea that both the z-spin-up and the x-spin-up preparations could possibly yield the same physical state S is sunk. The pair of electrons both in the state S would have to react somehow to our experiment, but all four possible outcomes have been forbidden by the quantum recipe.

The argument just given does not refute ψ-statistical or ψ-credal theories in their full generality. All we have shown is that electrons prepared z-spin up and electrons prepared x-spin up cannot possibly be in the same physical state. But what we would need to show is that the same holds for any pair of preparations that, according to the quantum recipe, yield different wavefunctions. One might even suspect that $|z\uparrow>$ and $|x\uparrow>$ preparations are quite likely candidates for yielding incompatible physical states since they are complementary: Maximal predictive certainty about z-spin implies minimal predictive certainty about x-spin, and vice versa. It is much more plausible that a preparation yielding $|z\uparrow>$ and a preparation yielding a more nearby wavefunction, such as $|30°\uparrow>$, might create individual electrons in the same physical state.

Pusey, Barrett, and Rudolph prove that this is not so. To rule out the possibility for such nearby states, one has to go to more trouble than for $|z\uparrow>$ and $|x\uparrow>$. For example, one might have to consider creating four electrons rather than just two, with each electron either prepared to be $|z\uparrow>$ or $|30°\uparrow>$. Now there are 16 possible preparations (each electron could be prepared either way), and we consider an experiment with 16 possible outcomes and show that each outcome is inconsistent with one possible preparation. In this way, PBR show that preparations yielding different wavefunctions can never yield electrons in the very same physical state. Hence the wavefunction of an individual electron does reflect some objective physical aspect of its physical state, and electrons ascribed different wavefunctions cannot be physically identical. The ψ-statistical and ψ-credal approaches to the wavefunction are incompatible with the predictions of the quantum recipe: Some ψ-ontic approach must be correct. The wavefunction somehow represents a real physical aspect of individual systems.

Our quest to understand how the physical world may be, given the predictive success of the quantum recipe, has made real progress due to two theorems: Bell's theorem and the PBR theorem. Bell teaches us that the predictions cannot be recovered by any local theory, that is, any theory according to which each of a pair of widely separated particles always has its own physical state that regulates its physical behavior (either deterministically or probabilistically) independently of what happens to the other particle. PBR teaches us that we must take the mathematical wavefunction assigned to an individual system as reflective of some real aspect of the physical state of the system.

In the name of completeness, we must mention that both Bell's theorem and the PBR theorem rely on one more premise, a premise so natural and universal that it is hard to imagine how to plausibly deny it. That premise, which is sometimes called "statistical independence," requires that there exist methods of "randomly" selecting either the setting of measurement apparatuses (Bell) or the particular prepared systems to experiment on from collections of prepared systems (PBR). The randomness entails that the selected groups will be not be correlated with the state of the incoming particles (Bell) or with the nature of the forthcoming experiment (PBR). Since the selection may be made however we like—flipping a coin, selecting on the basis of the parity of the digits, and so forth—to deny this premise is to assert the existence of some deep-seated conspiracy that would undermine all of experimental practice. (See the discussion in chapter 12 of Bell 2004).

Both the PBR and Bell theorems are "no-go" results: they tell us that certain approaches to understanding the physical world cannot succeed if the quantum recipe is accurate. But they both still leave quite a lot of room for articulating concrete theories. The nonlocality implied by the violation of Bell's inequality must be accounted for somehow. The wavefunction must represent some aspect of the real physical situation of individual systems. But there are alternative ways to implement the nonlocality and alternative accounts of the quantum state that the wavefunction represents. The detailed, concrete theories that we will examine illustrate some of these possibilities.

What Is the Quantum State after All?

Having established that the wavefunction of an individual system does represent some real physical aspect or feature of the system, we are naturally led to ask what sort of aspect that might be. If by stipulation we call this aspect the "quantum state of the system," our question then becomes: What is a quantum state?

It is not immediately clear what sort of question is being asked here. What information are we seeking? What characterizations of the quantum state might we expect to be enlightening?

One way of pursuing this question leads to a dead end. In a philosophical setting, one can be tempted to ask: to what *category of being* should the quantum state be assigned? The notion that all existing entities belong in one category or other goes back at least to Aristotle, who provided a canonical set of categories: substance, quantity, quality, relations, actions, passions, and the like.[5] Asking what a quantum state is, then, is asking to pigeonhole it in one or another of a set of pre-established ontological boxes.

I can see no virtue in this method of pursuing the question. Why think that Aristotle, or any other philosopher or scientist who never considered quantum theory, had developed the right conceptual categories for characterizing everything physically real? The quantum state is a novel feature of reality on any view, and there is nothing wrong with allowing it a novel category: *quantum state*. This is, of course, not an informative thing to say, but it does free us from the misguided desire to liken the quantum state to anything we are already familiar with.

[5] Aristotle *Categories* 1b25–2a4.

What sort of substantive questions about the nature of the quantum state can we ask, if the question of its ontological category is uninteresting? There are lots of detailed questions available. The main ones concern the relation between the wavefunction—the mathematical item used in the quantum recipe—and the quantum state. For example: Which mathematical degrees of freedom in the wavefunction correspond to physical degrees of freedom in the quantum state, and which do not? We have already noted that multiplying a wavefunction by an arbitrary overall phase does not change any predictions produced by the quantum recipe. This suggests that there is no physical degree of freedom in the quantum state that corresponds to the overall phase of the wavefunction.

We can ask whether there are many distinct fundamental quantum states pertaining to the physical universe or only one. The entanglement of the wavefunction plays an important role in addressing this question. When the wavefunction of two systems is entangled, it cannot be represented mathematically as the product of separate wavefunctions for each system. In this sense, neither subsystem individually has its own wavefunction, but only the joint system does. Following this line of thought and taking account of the ubiquitous interactions that result in entanglement, one might conclude that only the entire physical universe as a whole has a wavefunction, and hence a quantum state.

Or was that argument too quick? It is true that when the wavefunction of two systems is entangled, it can't be expressed as the product of separate wavefunctions for each subsystem. But a different mathematical object, called a *density matrix*, can be assigned to each subsystem. The density matrix can be used in a slightly modified version of the quantum recipe to derive empirical predictions for the subsystem.

Take, for example, a pair of electrons whose assigned wavefunction is the singlet state. Suppose that we wish to make predictions only concerning experiments carried out on one of the electrons. Can the quantum recipe be adjusted to assign some mathematical object to that electron alone and use it to generate the predictions?

We can't assign any spinor to our electron. Every single-particle spinor is an eigenstate for spin in some direction—that is, it predicts with certainty how the electron will behave if a Stern-Gerlach magnet is oriented in that direction. But in the singlet state, no prediction with certainty can be made in any direction: For all possible directions, the quantum recipe assigns even odds to how an experiment will come out. But the slightly different sort of mathematical object known as a density matrix can be used to generate just this prediction.

A density matrix is in many ways similar to a weighted collection of wavefunctions. Suppose, for example, we have a source that produces *z*-spin-up electrons half the time and *z*-spin-down electrons half the time. If all we know about a particular electron is that it comes from that source, we can make predictions by equally weighting by .5 what the recipe predicts for *z*-spin up and for *z*-spin down. This obviously leads to a 50-50 prediction about the result of a *z*-spin experiment and also obviously 50-50 for an *x*-spin experiment (because each eigenstate of *z*-spin yields that prediction). And not so obviously, it also leads to the same 50-50 prediction for a Stern-Gerlach magnet oriented in any direction.

A density matrix is not exactly the same as a weighted collection of wavefunctions, since different weighted sums yield the same density matrix. For example, an equally weighted sum of *x*-spin up and *x*-spin down yields all the same predictions as the equally weighted sum of *z*-spin up and *z*-spin down, and hence corresponds to the same density matrix. So even though we cannot assign a wavefunction to each individual electron in a singlet state, we can assign to each a density matrix. Can we then regard the wavefunction assigned to the joint system as somehow deriving from the density matrices assigned to the parts?

We cannot. The problem is that the pair of density matrices fails to contain any information about correlations between experiments done on the two systems. In a singlet state, for example, we not only predict a 50% chance of an up outcome to a *z*-spin experiment done on each electron, we also predict a 0% chance that both results will be up or both down. Our ability to predict the strict anticorrelation between results (even though we can't predict the individual results) is exactly what led to the EPR argument for the incompleteness of the wavefunction as a representation of the physical state of the particles.

Here's another way to see this point. The singlet state has the form $\frac{1}{\sqrt{2}}|z\uparrow>_A|z\downarrow>_B - \frac{1}{\sqrt{2}}|z\downarrow>_A|z\uparrow>_B$. There is a similar-looking state called the $m = 0$ *triplet state* whose form is $\frac{1}{\sqrt{2}}|z\uparrow>_A|z\downarrow>_B + \frac{1}{\sqrt{2}}|z\downarrow>_A|z\uparrow>_B$ (this is exactly the state α mentioned in the previous section). The density matrix assigned to each electron in the singlet state is identical to the density matrix assigned to that electron in the $m=0$ triplet state. But the two wavefunctions yield different predictions for the pair of electrons. In particular, when x-spin experiments are done on both electrons, the singlet state predicts they will certainly have different outcomes, and the $m=0$ triplet state predicts they will both have the same outcome. (You can check! Homework problem!) So the density matrices assigned to the parts do not determine the wavefunction assigned to the whole. There is more information in the global wavefunction than in the collection of mathematical objects assigned to its subsystems.

In the context of the quantum recipe, the mathematics of the wavefunction suggests that the quantum state (whatever it is) is a fundamentally *global* sort of thing. The quantum state of a system is more—in a very concrete sense—than any collection of states that can be ascribed to its individual parts. Pursuing this line of thought in the obvious way, conjoining all the parts of the universe into a single system, suggests that ultimately there is only one fundamental quantum state: the quantum state of the entire universe. This somehow influences the behavior of all the parts of the universe, but (unlike in the old mechanical picture of the universe) the global behavior cannot be accounted for as just the sum of interactions among the individually specifiable parts. Insofar as we can attribute wavefunctions to individual proper subsystems of the universe, we have the right to wonder why and how they inherit their wavefunctions from the universal one.

This observation leads to another puzzle. If the only fundamental quantum state is the quantum state of the entire universe, how does quantum physics manage to succeed as a practical method for making predictions? After all, no one ever has known or written down or calculated with a wavefunction for the entire universe! And if the only fundamental quantum state is the quantum state of the entire universe, how can we get any clues from the quantum recipe about how it behaves? For example, we would like to know whether the quantum state of the universe changes with time or is static. But if we cannot be sure about what the wavefunction of the universe is, how can we even begin to guess how it might change with time?

In sum, there are many interesting questions to ask about the quantum state that avoid entirely the question of what "category of being" it falls in. We can ask fundamentally how many quantum states there are. We can ask which mathematical features of a wavefunction correspond to physical features of the quantum state. We can ask why the wavefunction has the mathematical form it does. We can ask what physical role the quantum state plays in producing observable behavior. We can ask whether—and how—the fundamental quantum state changes with time.

Different theories answer these questions different ways. Some answer them more clearly and straightforwardly than others. One touchstone to use in assessing the conceptual clarity of a proposed physical theory is how it addresses these sorts of questions. Indeed, we will use claims about the behavior of the quantum state as the first characteristic that separates different exact quantum theories from one another. This detailed investigation begins in the next chapter.

Problem

1) Verify that the $m=0$ triplet state has the property claimed above, that is, that it predicts perfect correlation rather than perfect anticorrelation when x-spin experiments are done on both sides. This is just a matter of replacing the z-spin eigenstates with their equivalents written in terms of x-spin eigenstates.

Further Reading

Various analyses of the status of the quantum state can be found in the book edited by Alyssa Ney and David Albert (2013).

CHAPTER 4

Collapse Theories and the Problem of Local Beables

WE ARE NOW in a position to begin our main project: discussion of precisely articulated physical theories that can account for the predictive accuracy of the quantum recipe. We want to set a high standard for ontological and dynamical precision and clarity. It should be clear exactly what each theory postulates to exist and clear exactly how that ontology is postulated to behave. In particular, the account of the behavior of things, the *dynamics* of the theory, should be presented in the form of equations. These might be deterministic equations, so that an initial state of a system can evolve in only one way, or equations for a stochastic process, where the same initial state can evolve in various different ways. In the latter case, the dynamical law should specify both which evolutions might take place and what the chances are for each of the possibilities.

The obvious place to begin the search for such precise physical theories is with the quantum recipe itself. The recipe associates a mathematical wavefunction with systems, provides an exact equation (Schrödinger's equation) for how that wavefunction evolves at least some of the time, and offers a (vaguely specified) set of directions for extracting probabilistic predictions from it. The PBR theorem has assured us that the wavefunction reflects something about the real physical state of the individual system. So one natural thing to do is to postulate a real physical item—the quantum state—that is represented by the wavefunction and then try to specify its dynamics in a way that directly mimics the quantum recipe.

The main difficulty for such an approach lies in the third part of the recipe: the invocation of Born's Rule. Born's Rule is unexpected, puzzling, and problematic for several reasons.

One reason is that nothing in the recipe until that point suggests anything probabilistic. Wavefunctions are assigned to systems—at least in certain experimental situations—in a regimented way. We have no choice, for example, about which wavefunction to assign to an electron in a cathode ray tube of specified design and voltage. All the electrons get assigned the same wavefunction. That wavefunction in turn evolves deterministically according to Schrödinger's equation. In a sense (treating the single-particle wavefunction as a field on physical space), it spreads out, interacts with both slits, develops interference bands, and so forth. There is nothing inherently probabilistic in this, as the parallel case of water waves illustrates. It is only at the last step of the recipe that probabilistic notions appear. All of a sudden, without warning, the squared amplitude of the wavefunction is regarded as a probability, and what it is a probability for is a localized sort of event: a flash or mark appearing at one location on the screen rather than another. The wavefunction, which was being used to represent the electron, becomes progressively more spread out in space. But the phenomenon associated with Born's Rule is not spread out: It is highly localized. So both the sudden injection of probability into the theory and the sharp localization of the phenomenon have to be accounted for by the physical theory.

Born's Rule is problematic for a third reason: It does not precisely specify in exactly which physical circumstances it applies. Nor does it precisely specify exactly what wavefunction should be associated with a system after one uses it. The recipe, as presented, does not address this issue. A proper physical theory should do better.

Certain words are used when explicating Born's Rule: "Use it when a measurement is made." In the form we have appealed to, one would say: "Use it when a *position* measurement is made on the system and use the squared amplitude to assign various probabilities to the outcome of the measurement." But these words do not provide a precise physical characterization of anything. We have the rough-and-ready idea that the interaction with the phosphorescent screen in our experiment should be regarded as a position "measurement," so Born's rule should be applied then and not earlier. But this rough-and-ready sense has not been given any foundation in clear physical language.

Most standard textbook presentations do not confine the use of Born's Rule to position "measurements." They invoke the rule for "measurements" of many different observables: momentum, angular momentum, *z*-spin, and the like. Each observable property is associated with a Hermitian operator, and when a measurement of the property occurs, the wavefunction collapses into an eigenstate of that operator. But since experimental arrangements do not come with signs saying what they are "measurements" of, this makes the situation even more dire. Without a determination of when a "measurement" occurs and of what is being measured, one knows neither when to apply to Born's Rule nor how to apply it. Nor does one know what the wavefunction after the application should be.

Here I have consistently put scare quotes around the word "measurement," because in everyday life, the term carries connotations that could easily be physically misleading. Not every experiment is a measurement in the everyday sense. Here is a specification of an experimental procedure: Take a US quarter, and using your thumb, flip it so it travels at least 3 feet in the vertical direction and rotates at least six times in the air, let it land on a hard, flat, wooden floor, and record whether it lands heads up or tails up. This is a well-defined (enough) experimental procedure, but it is not, in any interesting sense, the measurement of anything. In particular, it is not the measurement of any property of the coin.

If the procedure is repeated many times, then one can say that the net result measures the bias of the coin. If after 10,000 repetitions, the coin has landed heads 70% of the time, we would confidently conclude that the coin is not fair (ignoring the possibility that the flipper had preternatural control of the flipping process). The bias of the coin is therefore regarded as a measurable physical feature of it. The normal notion of "measurement" has exactly this form: A measurement is not just any old physical interaction with a system, but an interaction so designed as to yield information about *features the system had antecedently to the interaction.*

When I judge that I measure my weight by stepping on a scale, I presume that I *had* a weight before stepping on the scale, and further that the behavior of the scale is correlated with that preexisting weight. If either of these conditions fails, then the interaction, whatever it may be, is not a measurement of my weight.

To call the physical interaction between the electron and the screen that produces the mark a "position measurement" therefore suggests that 1) the electron had a position antecedent to the interaction, and 2) the location of the mark is a reliable indicator of that position. This is exactly the sort of claim that will be either vindicated or refuted by the complete physical account of the interaction provided by a clearly articulated physics. But it is not the sort of thing that can just be read off from the phenomena or the data. We know by observation that a mark was created. We do not know by observation whether it corresponds to an antecedently existing position of the electron.

Quantum mechanics is often said to have a conceptual problem called the "measurement problem." Various things have been meant by this statement. In one formulation, it is the demand that a properly formulated quantum theory account for the fact that measurements have outcomes, a standard example being the case of Schrödinger's cat. Schrödinger imagined a cat imprisoned in a device with a radioactive source coupled to machinery that will release poison if a Geiger detector clicks. If the radioactive source is weak enough, after an hour, the wavefunction of the system (calculated using Schrödinger's equation) will have become wildly spread out in the configuration space of the system. The wavefunction will be a superposition of many states corresponding to the Geiger counter having gone off at different times and so killing the cat, and a state in which it never went off and the cat is fine. But although the wavefunction is agnostic about the fate of the cat, the cat itself (we think!) simply ends up either alive or dead. Born's Rule can be used to calculate a probability for each outcome, but it does not specify how or when the fate of the cat was decided.

In this formulation, the measurement problem has nothing to do with measurements per se. It is rather the problem of physically explaining how experiments come to have the sorts of outcomes we take them to have. Whether the experiments are "measurements" is immaterial. In fact, it is not clear that Schrödinger's experimental arrangement measures anything—the point is that it is an arrangement that either kills the cat or doesn't. Similarly, our coin-flip experiment is not a measurement, but we think it has a unique outcome: heads or tails.

Our present complaint is different. We have been trying to use the quantum recipe as a guide to the construction of a precise physical theory, but the recipe employs Born's Rule, which is itself explicated in terms of "measurements" of "observables," such as position. At the moment, we have no idea how to translate these terms into plain physical language. Which physical characteristics of an experiment determine whether it is a situation in which Born's Rule ought to be invoked, and, if so, which "property" has been "observed"? We seem to be confronted with the monumental task of explicating the concept of observation in precise physical terms.

So it is a nontrivial task to formulate a precise physical theory that explains the success of the quantum recipe and also uses the recipe itself as a model. The problems mostly arise recovering Born's Rule. The rule tells one to introduce probabilistic terminology into the description, but it does not give precise conditions when this should be done. And the rule, in a case like our phosphorescent screen, tells us to expect outcomes that are sharply localized in space, even though the wavefunction—the only mathematical representation we have of our system—has no such corresponding localization. Still, one can take the bull by the horns and try to solve these problems.

Collapse Theories

Various distinct physical theories can validate the quantum recipe. It is useful to divide these theories into generic types, illustrating each type with a concrete example. Our first type is collapse or reduction theories. The hallmarks of a collapse theory are 1) it postulates a fundamental physical quantum state that does not always obey a deterministic linear law of motion (such as would be represented by a wavefunction that satisfies Schrödinger's equation), but instead at least sometimes obeys an indeterministic and nonlinear law, and 2) this fundamental quantum state is *informationally complete*, that is, any two systems that have exactly the same quantum state are physically alike in all respects. In Einstein's terminology, according to a collapse theory, God does play dice, and the quantum-mechanical description of a system is complete. (Since these are exactly the characteristics Einstein attributes to "present quantum theory," we can infer that he regarded the prevalent understanding of quantum theory to be a collapse theory.) We assume here that the wavefunction of a system specifies a unique quantum state for the system. This is consistent with the wavefunction having mathematical degrees of freedom that do not correspond to physical degrees of freedom in the quantum state (i.e., different wavefunctions might specify the same quantum state).[1] But so long as each wavefunction is consistent with only one quantum state, the wavefunction will also be informationally complete. That is, it is possible in principle to extract every physical fact about a system from its wavefunction.

This general characterization of a collapse theory leaves two places for specific details to be filled in: 1) what the exact dynamics for the quantum state is and 2) how the quantum state determines all physical facts (including the observable outcomes of experiments). Theories that answer either of these questions differently are different physical theories. Initially, most of the work on collapse theories was directed at the first question, and various alternative answers were developed. After a while, it became clear that the second question is equally important and must be addressed. Let's start with the first.

In the standard textbook account, the dynamics of the wavefunction has a dual character: smooth deterministic linear

[1]To take the obvious example, two wavefunctions that differ by an overall complex phase are standardly taken to represent the same quantum state.

Schrödinger evolution “when no measurement occurs,” and random probabilistic collapse to an eigenstate of the measurement set-up “when a measurement does occur.” But without a clear physical characterization of what a measurement is, or what property is being measured, this is of no use. The most prominent collapse theory to completely avoid this difficulty was proposed by GianCarlo Ghirardi, Alberto Rimini, and Tulio Weber (1986) and is known as the GRW theory. If one had been steeped in the standard approach, one might have thought that the GRW theory must offer an exact physical account of measurement as the physical trigger of collapse. But just the opposite is true: GRW is physically precise exactly because it completely ignores the notion of measurement. Collapses of the quantum state occur in this theory, but when and how they occur has nothing at all to do with what (if any) measurements happen to be taking place. The GRW collapses occur in a uniform way at randomly occurring times with a fixed probability per unit time for each particle in the universe. Since measurement is nowhere invoked in this dynamic, there is no need for a physical analysis of what a measurement is. The GRW theory cuts the Gordian knot of the measurement problem.

Because there is no environmental trigger for the collapse, GRW is called a *spontaneous* collapse theory.

Suppose, for example, we prepare a single electron to have an initial wavefunction that is an equal superposition of an electron traveling to the right and an electron traveling to the left. If we just shoot an electron off to the left, it would have a fairly precise momentum to the left and we would attribute to it a wavefunction |left>. This wavefunction has a lump in it, which (by Born's Rule) shows where a flash would likely occur on a screen if one were set up. This lump, by Schrödinger evolution, moves ever farther off to the left. Figure 18a depicts such a wavepacket. The packet has a fairly well-defined position, given by the bell-curve envelope, and also a fairly well-defined momentum, given by the wavelength of the curve inside the envelope. Neither the position nor the momentum is perfectly well defined (i.e., the wavefunction is not an eigenstate of either), because by the Heisenberg uncertainty

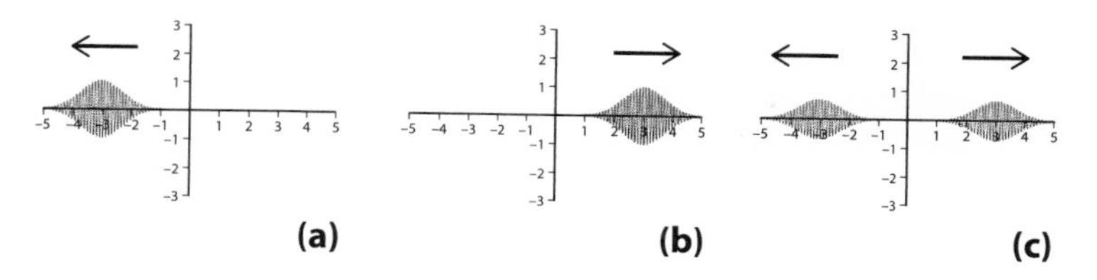

Figure 18

relations, if one is perfectly well defined then the other is completely unpredictable. Similarly, if we shoot an electron off to the right, we would attribute to it a wavefunction |right>, which is just the mirror image of |left>. Its lump moves ever farther to the right with little dispersion (Figure 18b). But then, by the superposition principle, there must exist an initial wavefunction that is $\frac{1}{\sqrt{2}}|\text{left}\rangle + \frac{1}{\sqrt{2}}|\text{right}\rangle$. So long as the dynamics of the wavefunction is given by Schrödinger's equation, this wavefunction must evolve with two equal-sized lumps moving off in opposite directions, getting farther and farther apart (Figure 18c). The perfect symmetry between the two sides will never change.

Since a spontaneous collapse theory (by our definition) is committed to the informational completeness of the quantum state, and hence the informational completeness of the wavefunction, this is already a somewhat curious situation. Where, we might ask, *is* the electron in this scenario? If the wavefunction is complete, then the electron is no more on the one side than on the other. One might be inclined to say that it is "smeared out" between the two locations in space, or that it is somehow in both locations, or even that it is really in neither location, but what one cannot possibly say is that such an electron is really on the right and not on the left.

Born's Rule tells us what to expect if we happen to put phosphorescent screens at some great distance on each side: there is a 50% chance of a flash occurring on the right, a 50% chance of a flash occurring on the left, a 0% chance of flashes occurring on both sides, and a 0% chance of no flash occurring on either side. But in this theoretical context, such an experiment cannot be regarded as a "measurement" in the intuitive sense, because the electron is not actually on the right rather than the left before the interaction with the screen, nor is it actually on the left rather than the right. So the flash cannot be interpreted as *revealing a preexistent fact about the electron*. In the proper sense of "measurement," this experiment is no more a measurement of any property of the electron than flipping a coin is a measurement of a feature of the coin. But Born's Rule, as a practical matter, does not require that any experiment actually be a measurement in this proper sense, merely that we would be inclined to *call it* a measurement, given a background in classical physics.

The important point is that whatever one might mean by "measurement," if the electron in our experiment just propagates in empty space, if no phosphorescent screen or other device is ever introduced into the situation, then no measurement of any sort can be considered to occur. Therefore, in any theory that associates the collapse of the wavefunction with "measurements," no collapse will ever occur: the wavefunction will always evolve linearly and symmetrically on both sides. But in the GRW theory, even if the electron is propagating in a vacuum, a collapse always has some chance of occurring, and if we wait long enough (a very long time, as we will see!) a collapse is essentially certain to occur. That collapse will break the symmetry of the wavefunction and will result in the particle being localized on one side of the experiment or the other.

How likely is such a collapse to occur according to the theory? This is determined by a new constant of nature, whose value can be set within a fairly wide range while recovering the verified predictive accuracy of the quantum recipe. In the original version of the theory, a single electron or proton would experience a collapse, on average, once every 10^{15} seconds, or about once every 10^{8} years.[2] So in this picture, starting out in the state $\frac{1}{\sqrt{2}}|left> +$ $\frac{1}{\sqrt{2}}|right>$ and evolving in empty space, the theory implies a tiny chance—about 1 in 1,000,000,000,000,000—that the quantum

[2] In some versions of the theory, the collapse rate depends on the mass of the particle and so would be different for electrons and protons. We ignore those subtleties here.

state of the system will collapse in any given second. So much for the timing of the collapse.

How does the wavefunction—and hence the quantum state—of the electron change if a collapse happens to occur? In the textbook account, this would depend on an analysis of the sort of measurement taking place, with the collapse resulting in a new wavefunction that is an eigenstate for that sort of measurement. Since the GRW collapses are completely detached from measurements, the answer must have an entirely different form.

In our discussion of the quantum recipe, we noted the incongruity that in many of our experiments the wavefunction tends to spread out and interfere over large areas, but the observed phenomena (flashes or marks) whose probabilities are provided by Born's Rule are relatively localized. Indeed, it is exactly because of this localization in the phenomena that one is inclined to regard interaction with the screen as providing a position measurement rather than, for example, a momentum measurement. The collapses of the quantum state in the GRW theory respect this idea, namely, that the collapse dynamics should counteract the tendency of the quantum state to spread out in space through Schrödinger evolution, having the opposite effect of making it bunch up. But just how bunched?

If one tries to model the collapse as the outcome of a position measurement, as it would be treated in the textbook approach, then the post-collapse state would have to be an eigenstate of the "position operator," that is, a wavefunction that yields a prediction with certainty about exactly where in space the particle will be found. But the only sort of state with the requisite predictive character is one in which the entire weight of the wavefunction is concentrated at a single point, with zero amplitude elsewhere. Properly speaking, there are no wavefunctions with this character (if a function is zero everywhere but at a single point, then its integral over any measureable set is also zero, and so it can't be normalized), but there is a fancier mathematical object called a *Dirac delta function* that has the required mathematical features. So perhaps after a collapse, the wavefunction should be such a delta function, reflecting the idea that the particle has become perfectly localized.

This idea fails on empirical grounds. We have seen how the Schrödinger equation works: The more rapidly the slope of the wavefunction varies in space in some direction, the faster (in time) the wavefunction spreads out in that direction. The slope of the delta function varies, as it were, infinitely quickly: It goes from being slope zero (horizontal) to an infinite slope (vertical) at a single point. So it would also spread infinitely fast. And this perfect localization would inject an enormous amount of energy into the system. If an electron in an atom were ever to suddenly have anything close to a delta function for a wavefunction, it would thereafter (according to the Schrödinger equation) be ionized and shot out of the atom with tremendous energy. We know that this is not happening, because we see no such high-energy electrons and do not notice any spontaneous increase of energy in physical systems.

The post-collapse localization of a particle, then, must not be such an extreme affair. In the GRW theory, the effect of a collapse is mathematically modeled by multiplying the wavefunction of the particle by a Gaussian—a bell curve—whose exact shape is described by another new constant of nature. To avoid the anomalous ionization problem, the Gaussian should not localize a particle into a volume smaller than an atom. In the original theory, the width of the Gaussian was taken to be 10^{-5} cm, or about 200 times the size of a hydrogen atom. (Once multiplied by the Gaussian, the wavefunction is renormalized.)

Having specified when collapses occur in the GRW theory—randomly, about once in 100,000,000 years for each particle—and what mathematical effect the collapse has on the wavefunction—multiplication by a Gaussian of width 10^{-5} cm—the theory still needs to specify where in space the Gaussian should be centered. A rough, but not quite mathematically accurate, prescription is that the probability density that the Gaussian be centered at a particular point in space is proportional to the squared amplitude of the pre-collapse wavefunction of the particle at that point.[3]

[3]What is mathematically accurate? The probability density that the Gaussian is centered at a point is not the absolute square of the pre-collapse wavefunction at that point, it is the convolution of the absolute square of the Gaussian with the

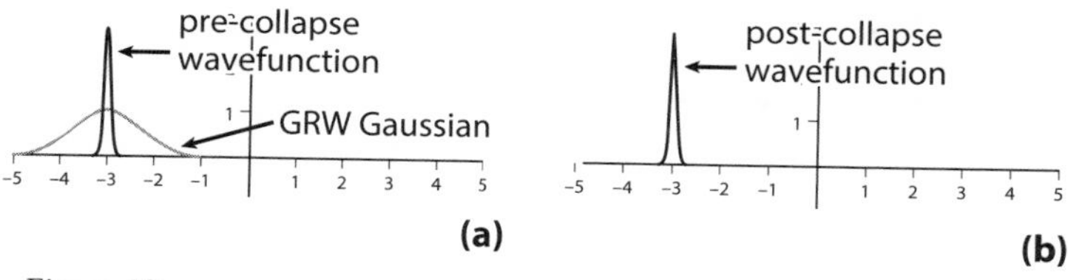

Figure 19

This connects the prescription for the probabilities for different locations of collapse directly to Born's Rule as applied to position measurements.

These precise mathematical prescriptions for the evolution of the wavefunction specify how the quantum state in the GRW theory behaves. The imprecise notion of "measurement" never occurs in the formulation of the theory. As far as the quantum state goes, everything else is just analysis.

Let's consider two examples of the effect of such a spontaneous collapse on the wavefunction of an electron. Figure 19a shows the wavefunction of an electron in an atomic orbital and the Gaussian that the wavefunction would be multiplied by if a collapse should occur. Since the collapse is only likely to occur where the squared amplitude is high, and the amplitude of the wavefunction at 100 times the atomic radius from the atom is essentially zero, the Gaussian will be centered somewhere over the atom. And since the width of the Gaussian is so much larger than that of the atom, the value of the Gaussian in the region over the atom is almost constant. So the waverfunction after the multiplication and renormalization will be almost exactly the same as before (Figure 19b). For all practical purposes, the wavefunction will continue to evolve in accord with Schrödinger's equation even if the electron happens to suffer a GRW collapse or "hit."

But the situation with our superposition $\frac{1}{\sqrt{2}}$|right> + $\frac{1}{\sqrt{2}}$|left> is quite different once the two lumps in the wavefunction have separated by even 1/100 cm. As Figure 20a shows, once the separation

---

absolute square of the pre-collapse wavefunction at that point. Aren't you glad we are skipping over these details?

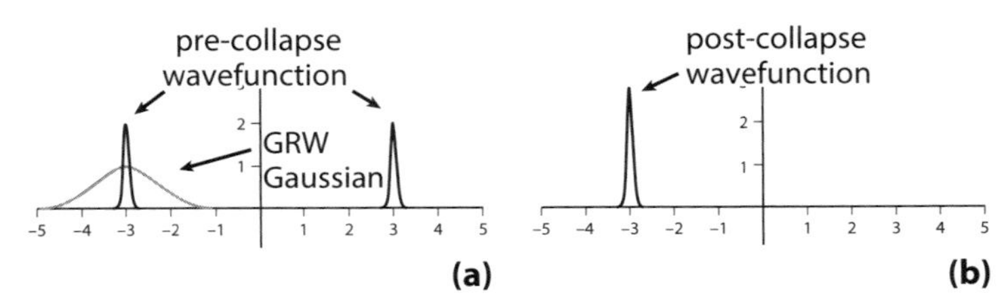

Figure 20

between the lumps is much greater than the width of the Gaussian, a hit will almost certainly be centered either over the right lump or over the left lump. Since the amplitude of the Gaussian rapidly approaches zero as it gets more than 10^{-5} cm from the center, the multiplication will reduce one lump to almost zero. The wavefunction goes from being symmetrically balanced between the two sides to being wildly unbalanced on one side or the other (Figure 20b). Either almost all the squared amplitude ends up on the right or almost all of it ends up on the left, depending on where the collapse is centered. And the equations imply that, in this case, there is a 50% chance of either outcome.

So under some circumstances, the GRW collapse has little effect on the wavefunction and in other circumstances, it makes a dramatic change. Still, one might wonder, what practical difference could any of this make? If each particle only suffers a collapse once in 100,000,000 years, how could that have any noticeable effect on the time scale of a laboratory experiment?

The key to answering this puzzle is entanglement. Consider an EPR set-up with a pair of electrons starting in the singlet state and both magnets oriented in the z-direction. Pure Schrödinger evolution yields the state

$$\frac{1}{\sqrt{2}}|z{\uparrow}>_R|\text{upward}>_R|\ z{\downarrow}>_L|\text{downward}>_L$$
$$-\frac{1}{\sqrt{2}}|z{\downarrow}>_R|\text{downward}>_R|z{\uparrow}>_L|\text{upward}>_L.$$

For the “up” and “down” outcomes of the experiment to be distinguishable, there must be much more than 10^{-5} cm difference

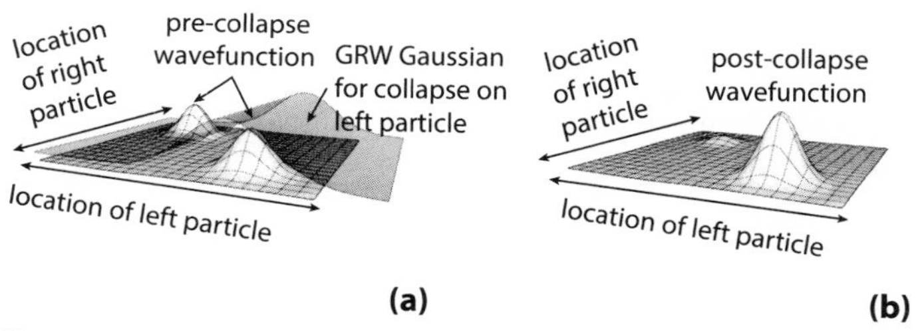

Figure 21

between where the upward and downward beams would hit a screen. So if a GRW collapse occurs on particle L, it will either accumulate almost all of the squared amplitude of the wavefunction in the $|\text{upward}\rangle_L$ branch of the wavefunction or in the $|\text{downward}\rangle_L$ branch, with equal chance of each outcome. But because of the entanglement in the wavefunction, such a GRW hit on particle L will also concentrate almost all the squared amplitude for particle R in the opposite branch to L. When particles are strongly entangled in the way they are in the wavefunction above, a collapse on either particle equally localizes both (Figure 21a, 21b). So the chance of both particles being localized in a given period of time is twice the chance of either individually suffering a collapse.

It is useful to compare the entangled state above with the product state

$$\left(\frac{1}{\sqrt{2}}\,|\,z\!\uparrow\rangle_R\,|\,\text{upward}\rangle_R + \frac{1}{\sqrt{2}}\,|\,z\!\downarrow\rangle_R\,|\,\text{downward}\rangle_R\right)$$
$$\left(\frac{1}{\sqrt{2}}\,|\,z\!\uparrow\rangle_L\,|\,\text{upward}\rangle_L + \frac{1}{\sqrt{2}}\,|\,z\!\downarrow\rangle_L\,|\,\text{downward}\rangle_L\right).$$

In the product state, the amplitudes of both R and L are equally divided between the upward and downward paths, so the chance of a flash occurring in each location, should a screen be introduced, is 50%. But as Figures 22a and Figure 22b illustrate, a collapse on one particle no longer has any effect on the wavefunction of the other. And if screens are introduced on both sides, there is no predicted correlation between the results. In 25% of the cases,

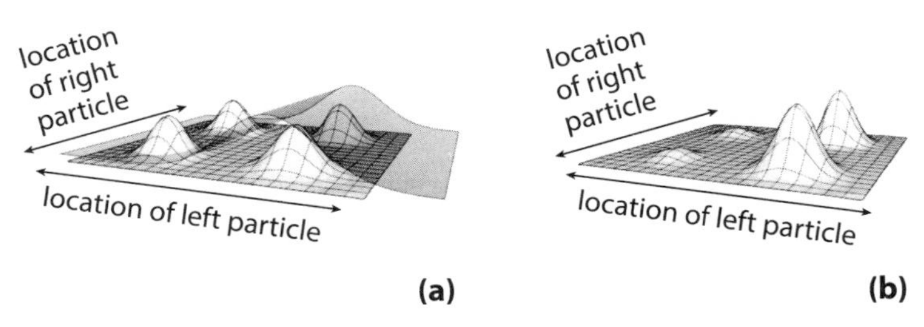

Figure 22

both flashes will be up, in 25% both down, in 25% R up and L down, and in 25% R down and L up. Unlike the singlet case, the location of the flash on one side provides no new information about where the flash on the other side might occur. In the product state, the localization of one particle does nothing to localize any others.

This effect scales with the number of particles entangled. When physical processes entangle all particles in a macroscopic object, the net effect becomes very significant. Suppose, for example, we start with a single $|x\uparrow>$ electron and pass it though a z-oriented magnet followed by a florescent screen. Next we add a light detector at the upper part of the screen, coupled to Schrödinger's infernal device that breaks a flask of poison if a flash is detected. Since a flash would result in the cat dying and no flash would result in the cat living, pure linear Schrödinger evolution of the wavefunction yields

$$\frac{1}{\sqrt{2}}|z\uparrow>|\text{upward}>|\text{dead cat}> - \frac{1}{\sqrt{2}}|z\downarrow>|\text{downward}>|\text{live cat}>.$$

The original electron is now entangled with every particle in the cat and, importantly, all particles in the cat are entangled with one another. This entanglement would also spread to the air molecules in the enclosure, since they would have different trajectories if the cat were breathing than if it were not. Quite plausibly, the spatial part of the wavefunction for every single particle in the apparatus differs by at least 10^{-5} cm. between the "dead cat" branch and the "live cat" branch of the wavefunction. So a GRW collapse on any one of these particles will have the effect of localizing all of them to one branch or the other.

And this, at last, is a result we can solve problems with. If we had to wait for a particular particle to suffer a GRW collapse in order for the wavefunction evolution to differ from pure Schrödinger evolution, then the GRW modification of the dynamics would make no difference over the time scales of actual experiments. But since there are more than 10^{27} particles in cat, a state like the one above could not survive the GRW dynamics for even a microsecond. Some particle will suffer a collapse in about 10^{-11} seconds, effectively eliminating one or the other branch of the wavefunction. Schrödinger's cat cannot remain in a state of indefinite health.

In short, the GRW dynamics, which is fully specified by precise equations of motion without any mention of "measurement," makes effectively no difference for the wavefunctions of single particles or small collections of particles on the time scale of centuries. Their wavefunctions will almost certainly obey pure linear evolution, thereby becoming delocalized. Their wavefunctions will interact with both slits in the Double Slit experiment and will travel both branches of an interferometer. But macroscopic solid objects, containing many particles bound to one another, cannot have wavefunctions that become delocalized in this way. A GRW collapse will almost instantaneously localize the wavefunction of each particle to a region of about 10^{-5} cm.

How does this new dynamics of the wavefunction help us understand the success of the quantum recipe? Granting one critical hypothesis, it is easy to finish the argument. The quantum state corresponding to

$$\frac{1}{\sqrt{2}}|z\uparrow>|\text{upward}>|\text{dead cat}> - \frac{1}{\sqrt{2}}|z\downarrow>|\text{downward}>|\text{live cat}>$$

is not, in this theoretical setting, one in which there is either a dead cat (and no live cat) or a live cat (and no dead cat). For we are assuming that the wavefunction is informationally complete, and the symmetry of the wavefunction ensures that it cannot represent the cat as being in one state rather than the other. But after a GRW collapse, the state will essentially be either

$$(1-\varepsilon)^{1/2}|z\uparrow\rangle|\text{upward}\rangle|\text{dead cat}\rangle -$$
$$(\varepsilon)^{1/2}|z\downarrow\rangle|\text{downward}\rangle|\text{live cat}\rangle$$

or

$$(\varepsilon)^{1/2}|z\uparrow\rangle|\text{upward}\rangle|\text{dead cat}\rangle -$$
$$(1-\varepsilon)^{1/2}|z\downarrow\rangle|\text{downward}\rangle|\text{live cat}\rangle$$

with ε extraordinarily close to zero. (ε never becomes zero, because the Gaussian never becomes zero, so neither branch is completely eliminated in a collapse. Whether this presents a problem should not be immediately clear.) The dynamics gives an equal chance to each outcome. So if we are justified in saying that the first state corresponds to the cat ending up dead (and not alive) and the second corresponds to the cat ending up alive (and not dead), then we will have vindicated the quantum recipe in this case.

In the same way, the GRW collapse dynamics will yield the same predictions as the quantum recipe, within observational error, for *every experiment that has ever actually been done.* It does not, however, yield the same predictions as the quantum recipe for all physically possible situations. We will return to this observation anon. We will also soon take up the critical question of whether we are yet justified in asserting that the first wavefunction above corresponds to a situation with a dead cat and the second to a situation with a live cat.

The Problem of Local Beables: Flash Ontology

Our discussion of the empirical consequences of the GRW theory has skated over one critical problem. We have, for example, simply assumed that if a particle in the system

$$\frac{1}{\sqrt{2}}|\text{right}\rangle_e|\text{detection}\rangle_d + \frac{1}{\sqrt{2}}|\text{left}\rangle_e|\text{no detection}\rangle_d$$

receives a GRW hit associated with the position of the particle in state $|\text{right}\rangle_e|\text{detection}\rangle_d$, then the empirical outcome is that of a pointer or gear moving one way, while if the hit is in a place associated with $|\text{left}>_e|\text{no detection}>_d$, then the pointer or gear doesn't move. But the GRW collapse is nothing but a change in the wavefunction or quantum state of a system, and the quantum state is itself not anything that exists in physical space. Since the pointer or gear is something in space, it is not immediately obvious how the behavior of the quantum state relates to the behavior of the pointer or gear. John Bell called this the problem of *local beables* in a theory. The term "beable" refers to items that exist according to the theory, things that are "just there." ("Beable" is pronounced bē-əbl, on the model of "observable." The "be" in "beable" is the English equivalent of the Greek "onta" in "ontology": The beables of a theory just are the ontology of the theory.) The quantum state is a beable of the GRW theory, but it is not a local beable, because it does not have values at points in space.

As we have seen, the wavefunction of a system is defined not on physical space but on a high-dimensional configuration space. The wavefunction has no value at a point in the three-dimensional physical space of the laboratory. Since the data of any theory are reported as the behavior of objects in space, a theory will be very difficult to interpret if it does not postulate anything existing in space, or indeed does not postulate any familiar space-time structure at all. And since we are assuming that the wavefunction is informationally complete, the motion and location of things in space must follow from it somehow. To give the theory empirical content, we need some sort of items that exist and move in physical space, influenced by the quantum state.

The GRW dynamics for the quantum state can be conjoined with several different postulates about local beables, thereby creating several different physical theories. Our rather glib talk of *the* GRW theory has been inaccurate. So far, I have presented only the GRW dynamics for the quantum state. We must now consider alternative ways to fill out the theory with local beables.

When John Bell presented the theory in his article "Are There Quantum Jumps?"[4] he was precise about the local beables he was

[4] Reprinted as Chapter 22 in Bell (2004).

positing. It is a surprising choice for the material occupants of space-time, which has since become known as the *flash ontology*. It is best to let Bell speak for himself:

> There is nothing in this theory but the wavefunction [Bell uses the term "theory" to refer to the mathematical formalism, as will become clear]. It is in the wavefunction that we must find an image of the physical world, and in particular of the arrangement of things in ordinary three-dimensional space. But the wavefunction as a whole lives in a much bigger space, of 3*N*-dimensions. It makes no sense to ask for the amplitude or phase or whatever of the wavefunction at a point in ordinary space. It has neither amplitude nor phase nor anything else until a multitude of points in ordinary three-space [i.e., an *N*-particle configuration] are specified. However, the GRW jumps (which are part of the wavefunction, not something else) are well localized in ordinary space. Indeed each is centered on a particular space-time point (**x**, *t*). So we can propose these events as the basis of the 'local beables' of the theory. These are the mathematical counterparts in the theory to real events at definite places and times in the real world (as distinct from the many purely mathematical constructions that occur in the working out of physical theories, as distinct from things that may be real but not localized, and as distinct from the 'observables' of other formulations of quantum mechanics, for which we have no use here). A piece of matter then is a galaxy of such events. As a schematic psychophysical parallelism we can suppose that our personal experience is more or less directly of events in particular pieces of matter, our brains, which events are in turn correlated with events in our bodies as a whole, and they in turn with events in the outer world.[5]

Bell's flash ontology proposes that the localized material content of space-time is not particles with continuous trajectories,

[5] Bell (2004), pp. 204–5.

nor continuously distributed fieldlike entities, nor vibrating strings, but rather *point events*. These points, whose locations correspond to the centers of the Gaussians of the GRW collapses, are scattered quite sparsely through space-time. Recall that an individual electron only suffers a GRW hit once in 10,000,000 years. So according to this theory, overwhelmingly most of the electrons associated with your body make no localized mark on space-time at all in the course of your entire life. They are all reflected in the wavefunction of your body, but nothing in physical space directly indicates their existence.

What, then, would this "galaxy of flashes" look like? Leaving the protons and neutrons (and hence quarks) aside, there are about 10^{28} electrons in a human body. That means, just from the electrons, about 10,000,000,000,000 flashes in a single second. The distribution of these flashes in space would trace out a quite detailed human form. There would be much less to this form in space than we commonly believe, but more than enough to define everything that we take to happen at macroscopic scale. The state of our brains (i.e., the flashes in our brains) would be reliably correlated with this distribution via the same physical mechanism, and thereby we can come to know what is happening around us.

But while the number and density of flashes is sufficient for a finely detailed spatial distribution at the macroscopic scale, it is quite paltry at a microscopic scale. There are about 40,000,000,000,000 cells in a human body, so (at the very rough scale of this estimate) only a few GRW collapses occur per cell per second. The distribution of flashes in space associated with a human body, then, would carry a lot of information about location, shape, and motion at macroscopic scale but almost nothing at the scale of individual cells. Of course, if we go about looking at a cell through even a regular optical microscope, the magnification process would entangle the wavefunction of the individual cell with macroscopic aspects of the equipment (or of the brain of the observer), and the "invisible" parts of the cell would automatically become registered in the GRW flashes.

This tremendous mismatch between what we think is going on spatially in a cell and what is going on according to this theory is extremely disconcerting. Indeed, the flash version of the GRW theory borders on a physical realization of a Cartesian demon, with the wavefunction playing the role of the deceiver orchestrating things. The physics predicts and explains the distribution of objects in space at the macroscopic scale in a way that renders our usual understanding of the goings-on at microscale completely wrong. But since ultimately all our empirical evidence for the theory must exist at the macroscopic scale, the theory still counts as empirically impeccable. Any revulsion toward it is conceptual rather than empirical.

At the start of our investigations, we remarked that physics is the science of matter in motion, that the theory of space-time structure provides the arena for the motion, and that quantum theory should provide the detailed structure of the matter inhabiting space-time. Up until now, we have discussed many experiments that quantum theory must account for, all described (of necessity) at the macroscopic scale. But if the behavior of matter at the macroscopic scale is nothing but the cumulative behavior of its microscopic parts, then a completed rigorous physics should specify what those microscopic local elements are and how they behave. The GRW flash theory is our first complete example of how such a physics might be constructed.

Many readers may be perplexed that the flash ontology, besides being rather odd and counterintuitive, bears no resemblance to anything he or she has ever come across in a physics text. I can only reiterate: That is not because a standard physics text provides an alternative account of what is going on at microscopic scale, but because it provides no account at all. In the standard textbook account, there is no local microscopic structure that can underpin macroscopic objects. Here, in contrast, we have a theory where everything is in the fundamental equations and exact specification of ontology. The equations are postulated to hold always and to describe a fundamentally stochastic physics. And the behavior of the quantum state is related to events in space-time via the flash ontology. There is no mention of "measurement" or "observation" or "observables" in the articulation of the theory.

The peculiar thinness of the local beables at the microscopic scale does not detract from these accomplishments.

The Problem of Local Beables: Matter Density Ontology

The problem of local beables arises acutely for the GRW collapse theory in part because the construction of the theory tightly follows the structure of the quantum recipe. The recipe, in turn, provides clear mathematical expression only for the wavefunction and its smooth linear Schrödinger evolution. GRW replaces the vague "assign these probabilities to the possible outcomes of measurement when a measurement occurs" instruction of the recipe with sharp mathematics. But that mathematics still only deals with the dynamics of the wavefunction and, hence, of the quantum state that it represents. Since the quantum state is not a local beable, all this precision has no logical consequences for what the local beables of the theory might be. The only interpretive constraint we have adopted is that the wavefunction be informationally complete, so that the distribution in space-time of whatever local beables we happen to postulate must be determined by it. But this interpretational constraint still leaves a lot of latitude in constructing a theory. Bell's flash ontology provides one way to complete the construction, using the discreteness and spatiotemporal location of the collapse centers as the key to the local beables. One might say that in the GRW theory, the collapses account for the "particlelike" aspect of the wave-particle duality, and the flash ontology makes use of this in postulating the local beables. But the discreteness of the collapses in time yields a corresponding sparseness of the flashes. Most of the time, there is literally nothing at all material that is localized in space-time.

It is equally natural, beginning with the wavefunction evolving by Schrödinger's equation, to be impressed by the wavelike aspects of the quantum state and to seek a correspondingly wavelike local beable. As we have remarked, the wavefunction of a single spinless particle is a complex function on physical space-time, so it is easy enough to associate it with a local beable. In such a picture, a single electron in a Double Slit experiment literally spreads out in space, with some of it passing through each slit and the two parts coming together later to interfere. One problem with this picture is accounting for the localized individual mark that eventually forms on the screen. But if we supplement the Schrödinger dynamics with the GRW collapses, this problem seems tractable: When a GRW hit occurs, the spread-out electron gathers itself up in one place, with almost all the density suddenly concentrated within 10^{-5} cm of the location where, in the flash ontology, a flash would have occurred. And as a substantial bonus, any other perfectly entangled electrons will experience a similar sudden contraction.

But so far, all this talk of the electron being spread out in space and then suddenly contracting in space is loose talk. The wavefunction of any system with more than one particle is not even mathematically a function over space: it is a function over configuration space. And similarly, the collapse and contraction are not in physical space but in the much higher-dimensional space. So it is not immediately obvious how to use the behavior of the wavefunction in a scheme for postulating a more wavelike local beable.

The problem, essentially, is that the structure of the wavefunction must somehow be projected down from configuration space into physical space, and then the theory must postulate a local beable corresponding to that projection. In such a theory, the GRW collapses will not correspond to the sudden creation of a pointlike local beable, as happens in the flash theory, but rather to a sudden change in the distribution of a continuously distributed beable.

In the nonrelativistic theory, there is an obvious and tempting way to define such a projection. The wavefunction is defined over configuration space, and the squared amplitude of the wavefunction forms, in the mathematical sense, a probability measure over the set of all possible configurations. If we ignore the suggestive word "probability" here, we can just say that the squared amplitude of the wavefunction defines a *weighting* of various configurations of particles. The natural suggestion, then, is to regard this weighting not as a probability but instead as a measure of how much of each particle is in each configuration. That is, if the squared amplitude of the wavefunction assigns a weight of .25 to a configuration in which a particular electron is on the left and a weight of .75 to a configuration in which that electron is on the right, then somehow .25 of the matter of the electron is on the left and the other .75 of the matter is on the right.

Since each possible configuration assigns an exact position to each particle, the weighting of the configurations can in this way be used to define a *matter distribution* for each particle. The matter of the particle literally gets smeared out over space. And as the wavefunction evolves in time, the matter distribution correspondingly evolves in time. This is a *matter density* local ontology of a GRW collapse theory.[6]

The local beables of the matter density theory have none of the peculiarities of the flash ontology. Every particle's matter always exists and is distributed some way in space. In a Double Slit experiment, the matter density of the electron literally spreads out and passes through each of the slits like a water wave. (In the flash ontology, by contrast, absolutely nothing exists in the space between the two slits and the screen, except on the very, very, very rare occasion when the electron suffers a collapse in transit.) That distribution will never be pointlike in any respect: The GRW collapses tend to concentrate the matter density in certain finite regions but never at a point. The matter density at the microscopic scale may be somewhat more spread out and amorphous than one would have thought, but it will certainly not be sparse. And at macroscopic scale, as with the flash ontology, the matter distribution will correspond to what we believe about where things are. Or at least it does so subject to some caveats.

The main caveat was noticed early on and goes by the name "the tails problem." The "tails" at issue are the infinitely extended, never-zero tails of the Gaussian that multiplies the wavefunction during a GRW hit. Since the tails are nowhere actually zero, the matter distribution after a hit is never driven to exactly zero

[6]For a more careful physical discussion of this sort of theory, see Ghirardi, Grassi, and Benatti (1995).

anywhere by the hit. To take a concrete example, if a Schrödinger-cat-like macroscopic superposition $\frac{1}{\sqrt{2}}|\text{right}>_e|\text{detection}>_d + \frac{1}{\sqrt{2}}|\text{left}>_e|\text{no detection}>_d$ suffers a GRW collapse, the collapse will drive one or the other components of the superposition almost to zero and therefore (given a matter density ontology) will concentrate almost all the matter into one configuration or the other. That is, the post-hit state will be something like

$$(1-\varepsilon)^{1/2}|\text{right}>|\text{detection}> - (\varepsilon)^{1/2}|\text{left}>|\text{no detection}>$$

$$\text{or } (\varepsilon)^{1/2}|\text{right}>|\text{detection}> - (1-\varepsilon)^{1/2}|\text{left}>|\text{no detection}>.$$

Subsequent hits, which are almost certain to occur on the initially favored configuration, will reduce the stray matter density exponentially lower. But for all that, the low density of extraneous matter will always exist. Should its existence concern us?

When these matters were first discussed in the philosophical literature, the problem of the tails seemed critical.[7] But that was because the basic interpretive principles used at that time were different from those we have been pursuing here. In particular, those discussions began with the textbook idea that Hermitian operators represent possible physical properties of a system and that the condition for a system to actually have the corresponding property is for its wavefunction to be an *eigenstate* of the operator. The value of the property would then be the corresponding eigenvalue. This interpretive principle is sometimes known as the *eigenstate-eigenvalue link*.

If one endorses this way of ascribing properties to physical systems, then the tails problem becomes acute, because a GRW collapse will not result in an eigenstate of the desired operator. If, for example, to have the property "the pointer having swung to right" a particular system must be in the state above labeled "|detection>," then the first post-hit wavefunction above does not describe a detector with it pointer swung to the right. The tiny bit of remaining entanglement prevents that state from having the desired property.

[7] See, for example, David Albert and Barry Loewer (1996).

The postulation of local beables cuts through this supposed difficulty. Our fundamental principle has been that the empirical consequences of a physical theory are determined by what the theory implies about the motion of matter, or better, about the distribution of local beables in space-time. For example, suppose that two theories have the same implications about what is in space-time and how it is distributed. Then it is hard to see how to maintain that they differ in their empirical content. If this is right, then the attribution of various properties to systems (e.g., the property of having x-spin up) does not, in itself, directly affect the empirical content of the theory unless the presence or absence of the property makes a difference to how matter will—or might—move. But in the GRW theory, the location of the local beables in space-time is completely determined by the quantum state, and the dynamics of the quantum state is probabilistically governed by the dynamical law. Any attempt to attribute "properties" to the system that go beyond specifying what the quantum state is and how the local beables depend on it would be idle. Nothing of significance rides on whether such additional properties exist. So it becomes irrelevant, in this theory, whether the wavefunction is an eigenstate of this or that operator.

Instead the question is whether there is any problem extracting a description of the macroscopic distribution of objects in space-time from the microscopic distribution. In the case of the flash ontology, no such problem presents itself. The only local beables associated with a system are the flashes associated with collapses on particles in the system, and the macroscopic location, shape, and motion of the object is nothing more than the collective distribution of those flashes in space-time. (Similarly, counterfactuals about how the object would or might have behaved under different circumstances are to be analyzed in terms of the probabilities of flash locations under the counterfactual supposition.) The further existence of tails to the wavefunction has no bearing on the actual spatiotemporal behavior of a system, given that no collapses actually occur in the location of the tails.

In the matter-density ontology, though, there really is something in space whose structure and behavior are determined by the behavior of the tails. The density of this matter is astonishingly small, but it is there nonetheless. At a first pass, for example, one might say that while most of the matter density when the quantum state is $(1-\varepsilon)^{1/2}$|right>|detection> – $(\varepsilon)^{1/2}$|left>|no detection> is in the shape of a detector whose pointer has (for example) swung to the right, there is still a tiny bit of matter density in the shape of a pointer that has not moved. (This first-pass description is, as David Wallace has noted, not technically correct: The way that the tails of the Gaussian fall off spatially implies that the matter distribution in the tail of the Gaussian will be distorted. Much more matter will be in the spatial direction closest to the collapse center.) It is tempting to advert to the smallness of this residual matter density as a reason to ignore it altogether, but the conceptual situation is not very straightforward. Here, for example, is a remark made by the physicist Philip Perle:

> In the GRW theory and the theory presented here [i.e., the theory discussed below] the state vector is never completely reduced. There is always a small but nonvanishing piece of "what might have been" included in the state vector. We do not regard this as satisfactory. If the reduced state vector is to correspond to what is actually observed in nature, it is hard to see what meaning can be given to an additional term that describes another observation, no matter how small.[8]

Perle's comments concern the state vector (i.e., the wavefunction) rather than the matter density, but the problem is obviously the same. Reduction of the density of matter does not eliminate it altogether, and if all it takes to make a physical object is to have some matter that behaves a certain way, then the GRW collapses with a matter density ontology seem to produce a Many Worlds theory, with a high- and a low-density Schrödinger cat at the end of the experiment.

If we do ignore the low-density matter, we can accept that the GRW collapse theory with a matter-density ontology makes the same predictions about the macroscopic behavior of objects as

[8] Perle (1989), p. 2289.

the GRW collapse theory with a flash ontology. Furthermore, the microscopic account is much less jarring. And as a further mark in its favor, the matter-density ontology is easily adaptable to modifications of the basic dynamics. For example, the discrete character of the GRW collapses can appear unnatural and ad hoc. This inspired research into continuous versions of the dynamics, known as Continuous Spontaneous Localization (CSL) theories. Pioneering work on such models was done by Perle (1989). In a CSL theory, the wavefunction evolves smoothly, albeit randomly. A given initial wavefunction can evolve different ways, with the basic dynamical equation providing the probability for each possible evolution.

Since no sudden "hits" or "jumps" occur in a CSL model, there is no obvious way to adapt a flash ontology to it. But the matter-density ontology works in exactly the same way, unaffected by the smoothing out of the dynamics. This observation reinforces my contention that GRW with a flash ontology and GRW with a matter-density ontology are really quite different physical theories despite their empirical equivalence.

The flash ontology and the matter-density ontology provide different solutions to the problem of local beables, resulting in different physical accounts of what exists in the world. But a strong contrarian line of argument maintains that all this focus on the local beables postulated by a theory is mistaken. We turn to this argument next.

The Problem of Local Beables: Emergence

Experiments and their outcomes are described in terms of the positions and motions of various sorts of macroscopic objects in a familiar space-time setting. To extract predictions about how experiments will (or might) come out from a description in the language of fundamental physics, then, one must be able to derive claims about the behavior of macroscopic bodies from the theoretical description. Bell observed that the postulation of local beables in a familiar sort of space-time solves this problem: The spatial behavior of the macroscopic objects is understood as just the collective behavior of the local beables associated with the system. At the microscopic scale, these might be postulated to be flashes, mass densities, particles, or something else. If one asks how we can come to know about this macroscopic behavior, the theory answers by showing how the states of our brains (including their local beables) can, according to the theory, become reliably correlated with these macroscopic shapes and motions.

In Bell's approach, a physical theory should postulate as fundamental features of the physical world both a macroscopically familiar space-time structure and some sort of localized physical items in that structure. But many physicists and philosophers have maintained that this need not be done. Rather, they claim, the physics can postulate a very different sort of fundamental physical ontology, from which the more familiar sort of description of the space-time structure and its local contents emerges. In the case of a theory like GRW, it is further claimed that the everyday macroscopic picture can emerge from a fundamental ontology that contains only the quantum state, with no local beables or familiar space-time at all. If this claim is correct, then the last two sections have been a waste of time: The behavior of the GRW quantum state alone suffices to ground the whole physics.

One of the most explicit proponents of this account of emergence for the GRW theory is David Albert.[9] Albert has described the sort of emergence he has in mind in some detail, so let us examine his account closely. As we will see, if sense can be made of the notion of emergence here, it might be of use to the Many Worlds theory as well. Albert's contention is that there is no need for—and indeed no advantage to—supplementing the GRW quantum state and its dynamics with any further ontology at all. All there is fundamentally, in Albert's version of GRW, is the quantum state evolving in accordance with a certain (stochastic) equation of motion. From this evolution a familiar world of objects located in a familiar low-dimensional space-time is supposed to emerge of metaphysical necessity. We do not postulate

[9] See Albert (2014), Chapters 6 and 7.

either a low-dimensional space-time or any of Bell's local beables in it, but rather recover them by metaphysical analysis.

Before examining Albert's arguments in detail, it may help to pose one obvious question. We have just finished discussing two different versions of the GRW theory. These theories agree about the dynamics of the quantum state: It is governed by the same equation in both cases. But they disagree about the particulars of the local beables, and this disagreement is quite substantial. So we might ask: If, according to Albert's account, some local beables in a familiar low-dimensional space-time emerge from just the dynamics of the quantum state, which local beables so emerge? Is it the flash ontology? The matter-density ontology? Both? Neither? What exactly is going on at the microscale in this picture?

Albert's reply[10] is that the question is not well posed. The notion of emergence he has in mind has an approximative character, and there can be emergent local objects in an emergent low-dimensional space-time at the macroscale but no emergent localized parts of them at fine levels of (emergent) microscale. So the relation between the spatial characteristics of these emergent objects and the spatial characteristics of their parts is not at all the same as the relation between macroscopic objects and flashes in the flash ontology or macroscopic objects and matter density in the matter-density ontology. In those theories, we postulate that some local beables exist with a sharp spatial distribution at all scales and that the facts at grosser scales are just the collective facts at finer scales described in a coarse-grained vocabulary. If this account of emergence is going to work at all, it will not leave us with anything like the relation between large objects and their smaller parts that we have been presuming.

Albert's strategy has two steps. The first is to argue that in any theory, the right way to metaphysically define the nonfundamental objects that emerge from the fundamental ontology is functionally. The second step is to argue that in the "bare" GRW theory (i.e., the GRW theory with no additional fundamental ontology besides the quantum state) there will be items that satisfy

[10] David Albert, personal communication, August 2014.

the relevant functional definitions. So in that theory, familiar low-dimensional space inhabited by macroscopic objects already exists without reference to any postulated fundamental low-dimensional space or fundamental local beables in that space.

What do the relevant functional definitions look like, and what, given the fundamental ontology of the theory, is supposed to satisfy these definitions?

At the level of common macro-objects, the answer to the first question is rather vague. We are told, for example, that "what it is to be a table or a chair or a building or a person is—at the end of the day—*to occupy a certain position in the causal map of the world*."[11] But what exactly a "causal map" is, or what a "position" in such a map is, or how something can "occupy a position" in the map, is not spelled out for these cases. How is one to go about specifying a "position in a causal map" for a table? (In contrast, there is no obvious problem specifying typical sorts of shapes and sizes of tables, how they typically move in space, and how they are related to one another and other things in space—just the sorts of things that can be easily read off a distribution of local beables in space-time.) Further, one of the typical causal contributions of tables is to keep things from falling to the floor, but a concept like "falling to the floor" seems much easier to characterize in terms of positions and motions than in terms of an abstract causal structure.

Even more puzzling in Albert's account is what is supposed to occupy these positions in causal maps. In one discussion of "emergent" particles in an "emergent" low dimensional space-time, what occupies the position are certain sets of coordinates. In a different context, mathematical projections of a single point particle in a high-dimensional space to lower-dimensional sub-spaces are recruited to do the job. Albert calls these mathematically defined projections "shadows." Think of the way the motion of a fly in a three-dimensional space can be projected down to the motion of the shadow of the fly on the two-dimensional floor, but leave aside any actual light casting a shadow or any floor: This

[11] Albert (2014), p. 127.

is a purely mathematical rather than physical operation. In the following passage, Albert considers an unfamiliar alternative to the classical Newtonian theory of many point particles moving in a low-dimensional space (and hence having a configuration that changes in time). In the alternative theory, there is only a single particle (the so-called world particle or "marvelous point") moving in a very high-dimensional space, of the same dimensionality as the configuration space of the normal theory. The critical argument is this:

> And if we pretend (just for the moment) that the laws of ordinary three-dimensional Newtonian mechanics, together with the three-dimensional Hamiltonian in equation (2), can accommodate the existence of the tables and chairs and baseballs of our everyday experience of the world—then we shall be able to speak (as well) of formal enactments of tables and chairs and baseballs, by which we will mean the projections of the position of the world particle onto tensor products of various of the $(3i\text{-}2, 3i\text{-}1, 3i)_C$ [these terms indicate three orthogonal directions in the high-dimensional space] subspaces of the *D*-dimensional space in which the world particle floats. And these formally enacted tables and chairs and baseballs are clearly going to have precisely the same causal relation to one another, and to their constituent formally enacted particles, as genuine tables and chairs and baseballs and their constituent particles do.
>
> And insofar (then) as we have anything in the neighborhood of a *functionalist* understanding of what it is to be a table or a chair or a baseball—insofar as *what it is* to be a table or a chair or a baseball can be captured in terms of *causal relations* of these objects to one another, and to their constituent particles, and so on, then these formally enacted tables and chairs and baseballs and particles must really *be* tables and chairs and baseballs and particles. And insofar as what it is to be a *sentient observer* can be captured in terms like these, then projections of the world particle onto those particular tensor products of three-dimensional

> subspaces of the *D*-dimensional space which *correspond* to such "observers" are necessarily going to have psychological experience. And it is plainly going to appear to such observers that the world is three-dimensional![12]

This argument provides the resources needed to demonstrate that the principles invoked here cannot be acceptable.

Consider a regular low-dimensional Newtonian world with tables and chairs and baseballs all composed of particles. And now define the "3-foot north projection" of any particle to be the point in space exactly three feet to the north (i.e. in the direction from the center of the earth to the center of Polaris) of the location of the particle. Then trivially the 3-foot north projections of all the particles in a table will be a set of locations that have the same geometrical structure as the particles in the table. And the 3-foot north projections of all the actual particles in tables and chair and baseballs will formally enact, in Albert's sense, the tables and chairs and baseballs and observers whose projections they are. But these "formal enactments" are clearly not tables and chairs, and the 3-foot north projection of a person having a headache is clearly not an actual sentient person with a headache. It might, in fact, just be a set of points in a vacuum (if the person is in a spaceship). But the 3-foot north projections in this world have all the same credentials—indeed even better credentials in terms of geometrical structure—as Albert's more abstract projections do. So Albert's argument cannot go through.

Where does it break down? One might contend that it breaks down at the claim that the projections stand in any causal relations to one another at all. And one might contend that it breaks down with the assumption that any of these sorts of macroscopic objects can be characterized as "positions in the causal map of the world." One might contend that it breaks down at both points. Individual readers might disagree about exactly where the flaw in the argument is. But on the assumption that no one will accept the existence of infinitely many numerically distinct, qualitatively

[12] Albert (2014), pp. 128–29.

identical tables and chairs and sentient observers (one for each N-foot north projection) in this scenario, then everyone will agree that the argument fails.

Recall Bell's description of what one is doing by postulating the flash ontology. The mathematical objects $(\mathbf{x}, t)$ are characterized by the mathematical structure of the wavefunction evolution: They are not physical at all. Each such mathematical object, given a coordinatization of space-time, corresponds to a physical point in a physical space-time. Those physical points exist independently of whether one postulates any flashes or not. What is needed now is the introduction of a novel *physical* entity: a point-like physical entity we have called a "flash." Flashes do not occur at every space-time point. The theory postulates both the existence of such things and characterizes (given an initial quantum state) a probability distribution over all the possible collections of flashes that might occur. A mere mathematical projection from the behavior of the quantum state down to the physical space-time does not make for any new physical ontology at all. In short, physical tables cannot be composed from mathematical projections. But mathematical projections can provide the resources to postulate the locations of physical entities, in a theory committed to such entities.

"Emergence" is a slippery word. Some forms of emergence—how an apparently continuous four-dimensional space-time could emerge from a discrete microscopic structure, or a seemingly continuous body of water could emerge from fundamentally molecular matter, or Newtonian gravity could emerge from the General Theory of Relativity in the limit of low relative velocities—are conceptually clear and straightforward. All that is meant by "emerge" here is that a literally false theoretical account (e.g., of water as a continuous fluid) can yield excellent approximations at certain scales. But how a low-dimensional space-time with particular macroscopic bodies could emerge from a fundamentally high-dimensional reality of the kind Albert postulates is a different matter. Albert has proposed a criterion for such emergence, but either the criterion itself is incorrect or the suggestion that it applies in the way he suggests to mathematical projections is. Any version of a spontaneous collapse theory that forgoes the postulation of local beables in a low-dimensional space-time faces substantial problems explaining how the physical world it describes could relate to the sort of experimental outcomes we are trying to explain.

The Experiments

In chapter 1, we discussed eight experiments, with the intention of using them as touchstones. One truly understands a proposed fundamental physical theory only if one understands how it would describe what is going on in these experiments and one understands how the theory could accurately predict the reported outcomes of the experiments. The experiments themselves are described at the macroscale, a level of description consistent with many different microscopic realities. Here, let us rehearse how these accounts would go for the GRW wavefunction dynamics with flash ontology and the GRW wavefunction dynamics with matter-density ontology.

In each of these theories, the fundamental physical description we have to work with includes a quantum state and distribution of local beables for the entire laboratory set-up. Any complete physical theory must be capable of treating the entire situation in terms of its postulated fundamental physical ontology. There can be no "measurement problem" or special axioms for measurements, since laboratory operations are treated like all other physical interactions. There will be a quantum state and local beables for everything in the lab, system and apparatus, observed and observer alike.

Let us presume that each of our experimental situations is described by some wavefunction for the entire laboratory set-up. We also assume that the experiment starts in a product state of a quantum state for the apparatus and a quantum state (supplied by the quantum recipe) for the electrons or other particles passing through the apparatus. Given such an initial state, the GRW dynamics will provide a probability measure for all possible collections of GRW collapses that might occur. And given the dependence of the location of local beables on the quantum state, that implies a probability measure over the possible distributions of the local beables. What we would like to show is that this probability measure makes it overwhelmingly likely that the outcomes of the experiments are what we take them to be, and what the quantum recipe predicts them to be. In the case of the matter-density version, we identify the outcome of an experiment with the behavior of nearly all of the matter density, ignoring the small quantity associated with the tails of the Gaussian.

Since the fundamental physical account has a distribution of local beables in space-time, it is obvious how to get from the microscopic description to the macroscopic and hence how to make contact with the language in which the experiment is described. Indeed, it seems clear how to do this using only the locations of the microscopic distribution of local beables as input. But there is no need to limit ourselves to just this information. Any information that can be derived from the fundamental physical description is available for use.

For example, the quantum state of our total system will be the sort of quantum state used to describe a system that contains electrons, protons, and neutrons (or electrons and quarks). So in the flash ontology or the matter-density ontology, the description of the distribution of local beables provided by the fundamental physics is not merely that there is some matter density or other in a particular location but that it is the matter density of an electron or of a proton. These different species are distinguished in the theory by constants, such as mass and charge, that appear in the wavefunctions and Hamiltonian of the system. In principle, then, we can extract the familiar macroscopic description not from just the distribution of local beables but from the characterization of the sort of particles whose local beables they are.

The simpler theory to treat in this way is the mass-density theory. Macroscopic equipment will appear in the fundamental account as a fairly stable mass density of electrons, protons, and neutrons, distributed so that the various atomic numbers of the materials would be evident at the microscopic scale. The cathode ray tube would be emitting a constant stream of electron mass-density (because the wavefunction of the system would be a superposition of individual electron emissions happening at different times). Some of this stream would reach the screen (Experiment 1: Cathode Ray Tube); be diffracted by the single slit and reach the screen (Experiment 2: Single Slit); or go through both slits, diffract, interfere into bands of greater and lesser density, and then reach the screen (Experiment 3: Double Slit). The complete physical description of the situation in every run of the experiment up to this point would be essentially identical (modulo random small contractions of the mass-density of the equipment due to collapses) no matter where the marks on the screen are later found. Indeed, the macroscopic distribution of mass density in each experiment will be exactly the same until some large system becomes entangled with the electron wavefunction in the right way. For example, a mechanical counter might be passed over the screen with an interaction Hamiltonian that would make the wavefunction evolve, by pure linear Schrödinger evolution, into a superposition of macroscopically different states, namely, having detected a mark at a particular place on the screen and not having detected a mark. Only at this point will the GRW collapses yield macroscopically different matter distributions. And the GRW dynamics implies that the probability of a collapse sending the matter distribution of the counter into the "detect" configuration rather than the "no-detect" configuration will be the probability assigned by the quantum recipe to getting a mark on the screen in that location.

Repeating the experiment over and over, by the law of large numbers, we eventually get a very high probability that the distribution of positions where marks are detected will be proportional to the squared amplitude of the wavefunction of the electron at that point on the screen. In this way, the theory predicts the reported data about interference bands in the Double Slit experiment.

Note that the varying distribution of matter-density of a single electron at the screen is not at all what is reported in the data. Each individual electron produces a single data point. Note also that the probabilities assigned in the theory to macroscopically different outcomes all derive from the fundamentally probabilistic evolution law. God does play dice in this theory: Experiments that start out physically identical will typically end up physically different. The fact that the matter density of a single electron displays the interference pattern it does at the screen is suggestive, but it is incidental to the empirical outcome. What accounts for the ultimate outcome is the interference in the quantum state and the amplitudes for different macroscopic outcomes that eventuate when the electron gets entangled with the macroscopic counter. It is at this point that the probabilities for the GRW collapses play their essential role, since the individual electron is fantastically unlikely to suffer any collapse during the experiment.

The interference bands go away in the Double Slit with Monitoring experiment, because the interference in the quantum state does disappear, as has already been described. Correspondingly, the matter density of a single electron at the screen will show no interference bands. The addition of spin to the quantum state makes no difference at all to the matter distribution until the spin degrees of freedom become entangled with spatial degrees. In the experiments using Stern-Gerlach magnets and the interferometer, the matter density of a single electron will get split and divided among the various paths through the device, except in the extremely unlikely event of a GRW hit in transit. If that event were to occur, almost all the matter density would suddenly become collected into one location along one path. The subsequent behavior would not show the usual interference effects, producing a slight deviation from the predictions of the quantum recipe.

Because a GRW hit causes a sudden global change in the quantum state, it also causes a sudden global change in the distribution of matter. Matter disappears from some locations and appears in others, which could be far away. So the behavior of matter density at the microscopic scale in this theory illustrates exactly the sort of jumpy correlated-over-long-distance behavior that Einstein discerned in the standard account of quantum theory. Even the EPR set-up, whose statistical predictions can in principle be recovered without any spooky action-at-a-distance, would display evident action-at-a-distance if one could directly see the matter density. God not only plays dice but also uses telepathic methods, just as Einstein claimed. In an EPR test with Stern-Gerlach magnets oriented in the same direction, for example, the physical situation in different runs of the experiment is essentially identical until a superposition of macroscopically different matter-density distributions starts to form on one side or the other. A GRW hit will soon shift most of the matter density to one outcome or the other, and the matter density on the other side will simultaneously change. It is through this coordinated global change of the physical state on both sides, no matter how distant, that the observed distant correlations are produced.

The same distant correlations produced by the collapses allow the theory to predict violations of Bell's inequality. That is, the collapse dynamics is not local in Bell's sense, so the constraints he demonstrates for local theories do not apply. If one could see the matter density shift around at the microscopic scale in the nonrelativistic setting, this nonlocality would be immediately obvious. And once again, this behavior of the microscopic matter distribution is not itself doing the heavy explanatory lifting: that is achieved by the coordinated global change in the quantum state, which the distribution of matter reflects.

In the mass-density theory, the reflection of the quantum state in the spatial matter distribution exists even for the low-amplitude parts of the quantum state, hence the tails problem. In this theory, the various possible outcomes of the experiment all continue to be inscribed in the actual behavior of matter in space, although at very different scales of density. Whether this difference in scale matters, though, is a subtle problem. If all the important concepts used to extract a macroscopic description from the microphysics are functional, then one could argue that the overall scale of a matter density is irrelevant to its functional characterization. If so, then both possible outcomes exist.

The flash theory, by contrast, does not suffer from this issue. At the level of what exists in space-time, there is no realization at all of one of the possible outcomes. The distribution of flashes corresponds to a unique course of macroscopic behavior. At the microscopic level, no matter "disappears" when a GRW collapse occurs; only a single flash appears. And there is no low-density ghost of the other possible outcome.

Of course, at the microscopic scale, astonishingly little is happening at all. Which raises a fascinating interpretive question: If the macroscopic spatial distribution of matter is determined by the microscopic distribution, just how detailed must the microscopic distribution be to constitute an acceptable physical theory? No sharp boundary exists here. In the flash theory, there is enough of a presence in space-time to unproblematically validate the sort of macroscopic descriptions found in the data reports and not nearly enough to validate the usual description of the microscopic structure of things. I have argued that this suffices to make the theory empirically adequate. Still, one might find it objectionable.

The EPR phenomena, as in the matter density theory, depend on the nonlocality implicit in the wavefunction collapse. At a mathematical level, this manifests itself in the probability structure of the predictions. Although there is a fundamental, irreducible 50% chance in an EPR *z*-spin set-up that the apparatus on the right displays an "up" result (by, say, the position of a pointer) and an irreducible 50% chance that the apparatus on the left displays an "up" result, there is zero chance that they both register "up" results. These chances are conditional on the complete physical state at the beginning of the experiment, and the probabilities derive from the indeterministic dynamics. The outcome on one side is not statistically or probabilistically independent from the outcome on the other, and the observed correlation is not the result of any predetermination in the initial conditions of the experiments. For in this theory, all experiments start out in exactly the same initial state. Still, the outcome on one side conveys information about the outcome on the other. Since this information is not already present in the initial state, there must be a real, physical dependency between the behaviors on both sides.

Since a nonlocality of this sort is employed to produce the results of the EPR experiment (where the phenomena themselves do not demand it), it is not too surprising that the theory can predict violations of Bell's inequality. The inequality only holds for theories whose physics is local and the GRW collapses (with their attendant consequences for behavior in space-time) are not.

The GRW theory, allied with either of these ontologies of local beables, has features that are not theoretically elegant. In the spontaneous collapse theory, the alternation between linear deterministic and nonlinear indeterministic evolution of the quantum state is jarring. This can be ameliorated in a CSL, although one would also lose the resources to specify a flash ontology. The appearance of new constants of nature quantifying the timing and spatial localization of the collapses is a surprise. And both ontologies of local beables appear somewhat problematic. GianCarlo Ghirardi himself remarks that the original theory was proposed as phenomenological rather than as a final, exact physics. But despite the somewhat provisional character of the theory, one should never lose sight of what has been achieved.

The GRW and CSL theories demonstrate that the sorts of phenomena described in our eight experiments—phenomena that display the types of effects most characteristic of quantum theory—can be accounted for by a physical theory devoid of any imprecision, unclarity, or obscurity in its physical postulates. The theory is articulated via sharp mathematics, with no mention of "observation," "observables," or "measurement." Different versions of the theory commit themselves to particular local beables, precisely distributed at microscopic scale, and the macroscopic behavior of the macroscopic objects mentioned in our experiments is understood as just the collective behavior of these microscopic facts.

To a great extent, the basic architecture of the GRW theory mimics the structure of the quantum recipe. But in place of the Born's Rule invocation of the vague concept of measurement, the GRW dynamics employ a sharp mathematical equation. The dynamics is universal: It obtains at all times and in all circumstances. There are no proposed modifications to logic or to probability theory; both of these topics are completely classical. And, in terms of this physical ontology and its exact dynamics, the predictions of the quantum recipe can be recovered. The claim that the phenomena of our experiments cannot be understood using classical logic or classical probability theory is therefore demonstrably false.

To be precise, the exact quantitative predictions of the standard recipe are approximately recovered. As noted earlier in the chapter, the collapses in theory must not narrow the wavefunction too extremely, as that would inject tremendous energy into the system, leading to spontaneous emission of electrons and spontaneous heating. Even the milder localization used in the theory has these consequences. If the GRW dynamics is correct, for example, the universe as a whole should be slightly warmer than the standard recipe predicts. Careful monitoring of samples of matter should show some anomalous emission of electrons. The scale of these effects is determined by the precise values of the new constants of nature introduced in the theory.

Physicists interested in collapse theories have calculated these effects and considered how experiments could verify or refute the theory. At this time, none of these effects have been seen. These null observations have put bounds on the possible values of the GRW constants but have not yet ruled out the approach. More exacting experiments may do so. A helpful review of the experimental situation can be found in Bassi and Ulbricht (2016).

Objective collapse theories of the quantum state are not the only sorts of theories that can achieve this conceptual and physical precision and clarity, but the GRW theory shows how this approach can be made to work. Other physicists who favor objective collapse theories have preferred to try to tie the collapses to a physical trigger rather than leaving them as a matter of pure chance. Roger Penrose, for example, has speculated that objective collapses are connected to gravity. The advantage of the GRW theory and its modifications and successors is that the speculative ideas are replaced by exact mathematics.

In the next chapter, we will see how a fundamentally different approach, with no collapse of the wavefunction, can also account for these same phenomena.

Further Reading

The GRW collapse theory is presented at various levels of technical detail in the following works: Albert (1992), chapter 5; Bell

(2004), chapter 22 (see also chapter 7 of this book for Bell's early discussion of the role of local beables in interpreting a physical theory); Ghirardi (2005), chapter 17; and Norsen (2017), chapter 9.

For a nontechnical sketch of a very different approach to wavefunction collapse, see Penrose (1989), pp. 367–71. For yet another approach, see Okon and Sudarsky (2014).

CHAPTER 5

Pilot Wave Theories

THE ARCHITECTURE OF COLLAPSE theories can be regarded as an attempt to follow the quantum recipe as closely as possible. A dynamics is specified for the quantum state that makes its behavior mirror that of the mathematical wavefunction used to make predictions. The "wave" (delocalized and interference) aspects of the phenomena arise from the linear evolution of the wavefunction, and the "particle" (highly localized) aspects from the collapses. But the wavefunction is not straightforwardly connected to physical space-time, nor is the quantum state. That connection is mediated via the local beables, whose distribution is determined by the quantum state. In this architecture, the exact nature of the local beables is the last problem addressed.

The architecture of a pilot wave theory is most easily grasped in the opposite order. First, one settles on the local beables of the theory. There are many options here. Given the local beables, it is already clear how the theory will generate predictions about the behavior of macroscopic objects in space-time: that will be determined by the collective behavior of the microscopic beables. This leaves consideration of the quantum state, and the wavefunction that represents it, as the last point of business. What effect or influence does the quantum state have on the local beables? Pursuing the questions in this order leaves the collapse of the wavefunction, as it occurs in the quantum recipe, as the very last thing to be explained.

As with collapse theories, we will focus on the nonrelativistic domain. Ultimately, of course, the nonrelativistic theory is not empirically accurate, both because relativity itself must be taken into account and because specific observable phenomena—such as particle creation and annihilation—are treated by relativistic quantum field theory rather than nonrelativistic quantum mechanics. But our first aim here is conceptual: How could one, in principle, go about constructing a precise physical theory capable of accounting for the characteristically quantum-mechanical phenomena in our eight experiments? Nonrelativistic quantum mechanics has an impressive roster of situations in which it makes extremely accurate predictions. Our programmatic assumption is that a theory that can reproduce those predictions provides a model of how a precise quantum theory can be constructed. Those basic structural ideas will have to be employed somewhat differently when dealing with relativistic quantum field theory. But the hope is that the general strategy of theory construction gleaned from the nonrelativistic context can be implemented.

This hope might be in vain. It is possible that only in the context of a relativistic treatment—or only in the context of accounting for particle creation and annihilation and other field-theoretic phenomena—do the critical physical principles arise. If so, our focus on the simpler case is leading us astray. But the advantage of starting with the simpler context is clear: The physics is more tractable and easier to grasp, and it has been worked out in great detail. With full cognizance of the assumptions we are making, let us press ahead.

The simplest and most familiar version of pilot wave theory has a very long history, going back nearly as far in time as quantum theory itself. It was presented by Louis de Broglie at the 1927 Solvay conference. The theory was not well received, and de Broglie soon stopped developing and defending it. It was rediscovered and amplified by David Bohm in 1952 and is therefore also sometimes called "Bohmian mechanics." Bohm presented the theory using Newtonian dynamics, supplemented with a so-called quantum potential derived from the wavefunction.[1] But a clearer and mathematically simpler presentation uses the *guidance equation*. This is how we will present it.

We begin not with the wavefunction or the quantum state it represents, but with the local beables. This is a theory of particles.

[1] See Bohm (1952).

The particles have definite positions at all times, and their positions evolve continuously in space. In the nonrelativistic theory, the particle number is fixed: Particles are neither created nor destroyed. They just move around, changing their configuration.

Since we start with particles in a familiar low-dimensional nonrelativistic space-time (that is, a space-time with a unique absolute time), defining the configuration of the system at a given time presents few difficulties, as we discussed in Chapter 2. Given the means to mathematically specify any configuration of the system, we can define *configuration space* as a mathematical space, each point of which uniquely represents a different possible configuration of the system, and which contains representations of all possible configurations. Configuration space is an abstract mathematical space, not a physical space.

It worthwhile to repeat here something mentioned in footnote 3 in Chapter 2: The mathematical structure of the configuration space of N distinguishable particles differs from that of N indistinguishable particles. A configuration of N distinguishable particles in a three-dimensional space is standardly represented by a point in R^{3N} (i.e., by an ordered $3N$-tuple of real numbers). Each successive triple of real numbers represents the position of one particular particle, with the numbers depending on how coordinates have been introduced into the space. Swapping the positions of two of the particles yields a different configuration, because the particles are distinguishable.

But if the N particles are indistinguishable, then switching particles does not yield a different configuration. The configuration of N indistinguishable particles in a three-dimensional space is just a set of N points taken from the space (assuming the particles cannot co-locate). The space of all sets of N points in, say, three-dimensional Euclidean space E^3 can be denoted by ${}^{N}E^3$. More generally, the N-identical-particle configuration space of a coordinatized three-dimensional space can be symbolized as ${}^{N}R^3$. This difference in mathematical representation of the configuration spaces of distinguishable and indistinguishable particles generates differences in the rest of the mathematical apparatus. These differences deserve some attention.

The physical question is exactly how these particles move according to the theory. So far, we have not mentioned any wavefunction or quantum state at all. All we have postulated as physical items are particles in a familiar four-dimensional classical space-time, whose motion is represented by the motion of a single point in the corresponding abstract $3N$-dimensional configuration space. But that abstract space is precisely the mathematical structure that is used to define the wavefunction of our system! The wavefunction assigns a complex number (or a complex spinor) to every point in this abstract space. The second ontological posit of our theory is a real, physically objective quantum state, which is represented (somehow) by this wavefunction.

How are we going to make use of the quantum state in specifying the dynamics of our particles? We want it to determine how the particles move, which is the same as determining how the configuration of particles changes. The changing configuration is represented by the motion of a single point in configuration space. So if we can mathematically specify how that point moves, we can represent how the configuration changes and hence how the particles move. Furthermore, if the particles move continuously and smoothly in the space, then the representative point will move continuously and smoothly in the configuration space. So to represent how the system of particles will behave, we want to specify how any given possible configuration of particles will change, which means specifying how any single point in the mathematical space will move, given where it is now. In short, what we want mathematically is a *velocity field on configuration space*. Such a velocity field associates a velocity with each point in configuration space, and that velocity determines how the particles in such a configuration would move.

From a purely mathematical standpoint, then, our problem is this: Given a wavefunction on configuration space, define a velocity field on configuration space. If we have that, then given an initial configuration and initial wavefunction (and given some dynamics for the wavefunction), everything is determined. These initial data fix a unique evolution of the system in time.

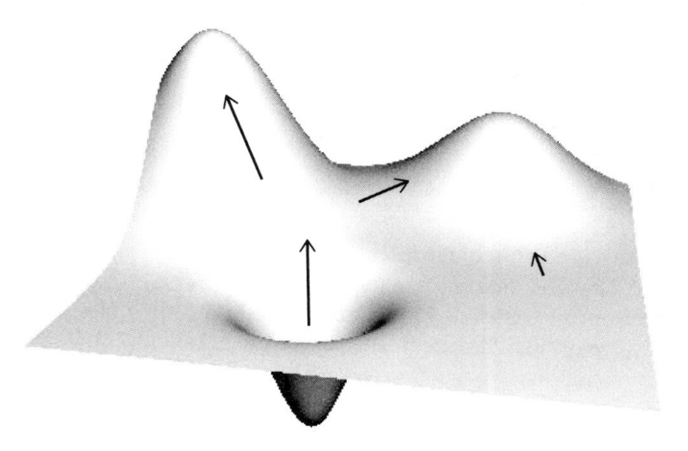

Figure 23

Leaving the spinors aside for the moment, the wavefunction is a *scalar field* on the configuration space: It assigns a single complex number to each point. And a velocity field is a *vector field*: It assigns a vector with a direction and magnitude to each point. So our mathematical question is: What is an obvious way to derive a vector field from a scalar field? The simplest mathematical answer is to take the gradient.

The gradient of a scalar field is a vector field that indicates how the value of the scalar is changing. Consider a familiar real number field—like a measure of the height of a hill at every point. The gradient at a point is an arrow pointing in the direction of steepest ascent from there, with the length of the arrow representing the slope (Figure 23).

Mathematically, the gradient is represented by the symbol ∇. Furthermore, we know we want any two wavefunctions that differ by a constant factor to represent the same quantum state, and hence to yield the same motion. A simple way to achieve this is to divide through by the wavefunction ψ: Multiplication in the numerator and denominator cancel out any constant factor. Since the wavefunction is a complex scalar field rather than a real scalar field, the result so far can be complex, but one way to deal with that is just to use the real or the imaginary part of the result. We choose the imaginary part. This approach yields (with some constants put in) this *guidance equation*:

$$\frac{dq_k}{dt} = \frac{\hbar}{m_k}\mathrm{Im}\left(\frac{\nabla_k \psi(x,t)}{\psi(x,t)}\right).$$

In this equation, q_k represents the position of the kth particle, so the left-hand side represents its velocity. The right-hand side, apart from the reduced Planck's constant ħ and the mass of the particle m_k, is just the imaginary part of the gradient of the wavefunction for the kth particle coordinates divided by the wavefunction. All wavefunctions belonging to the same ray in Hilbert space generate the same vector field, and so they are naturally understood as representing the same quantum state.

Our total theory, at this point, has the following structure. There is a classical space-time, with three spatial dimensions. The local beables in this space-time are N particles, which always have definite positions and move around. Each particle is characterized by a mass m_k. There is a single nonlocal beable, the quantum state, which is represented by the wavefunction ψ.

All we need to complete the theory is a dynamics for the wavefunction (and hence the quantum state). We postulate that the dynamics of the wavefunction is always described by Schrödinger's equation, which requires specification of a Hamiltonian operator $\widehat{H}$:

$$i\hbar\frac{\partial\psi(x,t)}{\partial t} = \widehat{H}\psi(x,t).$$

The wavefunction never collapses in this theory. If the potential term in the Hamiltonian requires it, the particles can also be characterized by physical quantities, such as electric charge or "color."

Since both Schrödinger's equation and the guidance equation are deterministic, the complete initial conditions for the physical world in this theory (i.e., an initial wavefunction and an initial configuration of particles) determine a unique evolution. There are no dynamical probabilities.

This dynamics is fundamentally unlike Newtonian mechanics. Newtonian mechanics is second-order in time; that is, the fundamental dynamical equation $\mathbf{F} = m\mathbf{A}$ specifies the accelerations of objects rather than their velocities. The initial conditions for a Newtonian problem are therefore the initial positions and initial velocities of all particles. In a certain sense, in Bohmian mechanics, the initial wavefunction replaces the initial velocities in Newton's theory. Via the guidance equation, the initial wavefunction determines the velocity at that time, just as it does at all times.

One might well wonder how the probabilistic predictions of the quantum recipe could possibly be consequences of this theory. Indeed, one might wonder how this theory relates to our experiments at all. But one thing is clear. What has just been specified is a theory of *something.*[2] Given an initial wavefunction and configuration, the particles will move in some definite way. What will that motion look like?

Figures 24a and 24b show different possible trajectories for an electron in the Single Slit and Double Slit experiments. We are assuming both the initial wavefunction and the Hamiltonian used standardly for these experiments. Since there is only a single particle being analyzed here (nothing else of note is moving in a significant way), the initial configuration is just the initial location of the electron. The initial position of a particle determines where it will intersect the screen, as shown by the trajectories.

How does one get from these trajectories to the predictions of the quantum recipe? Those predictions are statistical. They state what proportion of marks or flashes in different locations are likely to accumulate if the experiment is run many times. But unlike the GRW theory, where the initial state is just the wavefunction (and therefore the initial state is the same in all the runs), in Bohmian mechanics, the initial state comprises both the wavefunction and the initial particle location. Even if two experiments start with the same initial quantum state, they can have different outcomes due to different initial configurations. And we can only derive statistical predictions about the distribution of these

[2]Credit goes to Shelly Goldstein for this formulation.

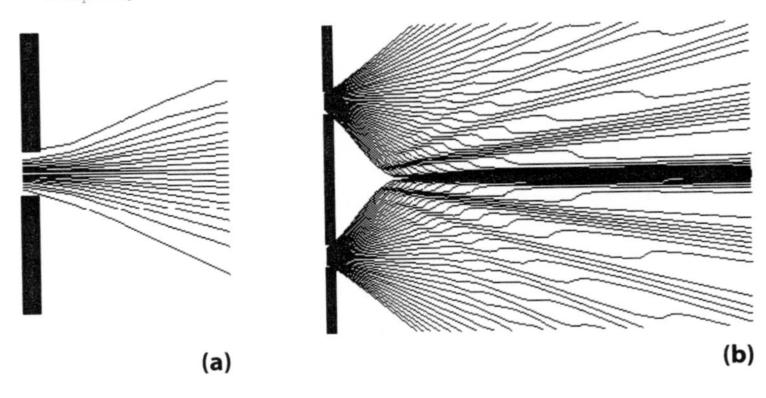

Figure 24

outcomes given a statistical hypothesis about the distribution of the initial locations.

For these particular experiments, the relevant statistical hypothesis is clear: Over many runs, the initial locations of the electrons are fairly evenly distributed over space in the beam before they reach the slits. Given such an initial distribution, we would predict, for example, that about the same number of electrons will go through each slit. We also predict that the number of particles in the Double Slit experiment that end up at a given part of the screen is roughly proportional to the initial region of space occupied by trajectories that reach that part. Given this assumption about the initial uniform distribution of electrons, it is now a purely mathematical fact that as more and more experiments are done, *the number of electrons reaching any given region of the screen is proportional to the squared amplitude of the wavefunction at the screen*. This fact counts as an explanation of the observed phenomena provided that 1) the hypothesis about the initial distribution is correct, and 2) the distribution of marks or flashes at the screen reflects the distribution of electrons that reach the screen.

In this particular case, the hypothesis about the initial distribution of particles is quite plausible on its own, just from the overall symmetries of the situation. Since the source is symmetrically related to the slits, there is no physical cause for more particles to arrive at one rather than the other slit in the long run. We can also check directly that the intensity of the beam before it reaches the slits is constant through the beam. Similarly, in the Double Slit case, we can trace the location of each interference band back to a "channel" that leads to it. Our only assumption is that the number of electrons entering each channel is proportional to the initial channel width. For that assumption to fail, there would have to be some physical influence accounting for a correlation between where the electrons start out and where the slits are located. The only assumption needed here to derive the predictions is the absence of any such correlation.

The form of the guidance equation guarantees a mathematically crucial property of the Bohmian trajectories, a property that helps explain the probabilistic consequences of the theory. We already know that the absolute square of the wavefunction yields a probability measure over configuration space (i.e., a non-negative measure whose integral is normalized to 1). As the wavefunction evolves in time, its absolute square evolves as well, yielding a *probability current*. One can think of the probability measure as flowing around in configuration space like an incompressible fluid. The trajectories for different actual configurations, as specified by the guidance equation, follow this probability current. Where the squared amplitude of the wavefunction tends to pile up in configuration space, the Bohmian trajectories tend to congregate as well and to exactly the same degree. We see this in Figure 24: Relatively more Bohmian trajectories end at exactly the points on the screen where the squared amplitude of the wavefunction is large, and few end where the squared amplitude is low.

As for justifying the claim that the observed marks will be located where the particles reach the screen, that calls for a complete physical account of the electron/screen interaction. It is via this account that Bohmian mechanics addresses the measurement problem. To get a handle on this account, we need to start thinking about interactions between particles. The Double Slit with Monitoring is our simplest example of such an interaction.

We described the set-up of that experiment as adding a proton in a chamber between the slits (see chapter 1). The idea is that if the electron goes through the upper slit, the proton will move upward; if it goes through the lower slit, the proton will move downward. How do we get such an interaction between the particles into the physics?

The only place available is in the potential term of the Hamiltonian. If there is no interaction potential and the initial wavefunction of the electron is not entangled with that of the proton, then the wavefunction will never become entangled. In such a case, the location of one particle will provide no information about the trajectory of the other: There is no monitoring. But with the right sort of interparticle potential, the wavefunction will become entangled, and the motions of the particles will become correlated. We can use the standard electrostatic potential energy from classical physics as a simple (indeed, slightly oversimplified) example.

The classical electrostatic potential energy is

$$\mathrm{U} = \frac{KQq}{|\vec{r}|},$$

Where Q is the charge on one particle, q the charge of the other, K is a constant, and r the vector representing the spatial separation between the particles. We can write the electric potential energy as $\mathrm{U} = \frac{Kq_e q_p}{|x_e - x_p|}$, where x_e represents the spatial coordinate(s) of the electron, and x_p the spatial coordinate(s) of the proton. We turn this classical expression into an operator by "putting on the hats," that is, replacing the classical position variable with a wavefunction operator. In addition to this interaction potential energy, there are terms in the Hamiltonian corresponding to the classical kinetic energy of the particles, yielding the total Hamiltonian:

$$\widehat{H} = \frac{-\hbar^2}{2m_e}\frac{\partial^2}{\partial x_e} + \frac{-\hbar^2}{2m_p}\frac{\partial^2}{\partial x_p} + \frac{Kq_e q_p}{|x_e - x_p|}.$$

The effect of the interaction term is to entangle the initially unentangled electron and proton wavefunctions. The guidance equation applied to this entangled wavefunction yields the configuration-space trajectories, two views of which are shown in Figure 25. Note that when the electron goes through the upper slit, the proton moves up, and when it goes through the lower slit,

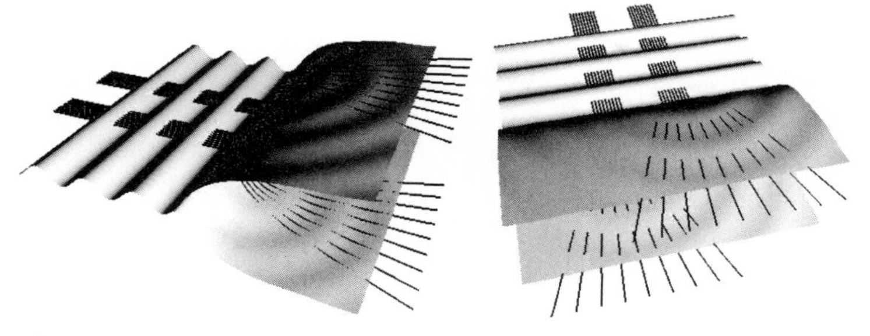

Figure 25

the proton moves down. In this circumstance, the wavefunction decoheres: The trajectory of an electron that goes through one slit is thereafter unaffected by the part of the wavefunction associated with the other slit, since the proton position causes these parts of the wavefunction to separate in configuration space. Assuming that the actual marks on the screen form where the electrons meet it and that the initial distribution of electrons (over many runs) is uniform in space, the theory predicts exactly what the quantum recipe predicts.

Decoherence already explains why there can be an "effective" collapse of the wavefunction in this theory, even though the wavefunction never actually collapses. The dynamics of the theory as represented in configuration space is local. That is, the change in the actual configuration of the particles is determined entirely by the value of the wavefunction near the corresponding point in configuration space (which determines the gradient). So any part of the wavefunction far from the actual configuration will have no immediate effect on how the particles behave. If such a part never evolves in the future to reach the region of configuration space where the actual configuration is, then that part can be ignored or thrown away without empirical consequence. The "collapse of the wavefunction" does just that.

We can also now see how to ensure that the marks on the screen will form (according to the theory) where the particle arrives. For the state of the screen to reflect the arrival location of the electrons, there must be an interaction between the two. And if that interaction is short range (like the one discussed above, which is proportional to $\frac{1}{x_e - x_p}$ and so gets small as the distance between the interacting particles gets large), then the only particles in the screen whose positions will be changed by interaction with the electron are particles in the screen *near* the electron. We can therefore see how changes in the configuration of particles in the screen will only occur where the electrons arrive. Just as the proton position can contain information about which slit the electron went through, the configuration of particles in the screen can contain information about where the electrons from the cathode ray tube arrive.

What about spin? One might think that just as one postulates an exact position for each particle at all times, so one should postulate an exact direction of spin, which evolves through time. But such a direction is never directly observable: What we record, using a Stern-Gerlach apparatus, is where the particle arrives on the screen. So what we really need is for the spin state, as represented in the wavefunction, to make a difference in how things move. To do that, the spin state has to somehow play a role in the guidance equation. There is a mathematically simple way to make the spin state affect where the particles go. Modify the guidance equation as follows:

$$\frac{dq_k}{dt} = \frac{\hbar}{m_k}\mathrm{Im}\left(\frac{\psi^*(x,t)\,\nabla_k\psi(x,t)}{\psi^*(x,t)\,\psi(x,t)}\right),$$

where $\psi^*(x, t)$ is the complex conjugate of $\psi(x, t)$.

On the surface, it looks like we have just multiplied the numerator and denominator by the same thing, which would have no effect at all. And for a wavefunction without spin, that is just what happens. But the situation is different with spinors. Recall that to get a velocity field, we want to take the gradient of a scalar function. Since spinors are not scalars, it is not immediately clear how to accommodate the spin variables in the guidance equation. But even though a spinor such as $\left[\begin{smallmatrix}a\\b\end{smallmatrix}\right]$ is not a scalar, $\left[\begin{smallmatrix}a\\b\end{smallmatrix}\right]^*$ is the same as $[a^* \;\; b^*]$, so $\left[\begin{smallmatrix}a\\b\end{smallmatrix}\right]^* \times \left[\begin{smallmatrix}c\\d\end{smallmatrix}\right] = [a^* \;\; b^*] \times \left[\begin{smallmatrix}c\\d\end{smallmatrix}\right] = a^*c + b^*d$, which is just a scalar (complex number). So this simple change has the effect of making the guidance equation apply to spinor wavefunctions just as well as it does to scalar wavefunctions.

In the Stern-Gerlach situation, the electromagnetic interaction added to the Hamiltonian creates entanglement between the spinoral degrees of freedom in the wavefunction and the spatial degrees of freedom. Given the guidance equation, this entanglement influences the particle trajectories. And again, as a purely mathematical consequence, the proportion of particles deflected up or down in this situation will match the predictions of the quantum recipe, assuming only that the initial distribution of the particles over many runs is approximately the squared amplitude of the initial wavefunction.

More detailed discussion of what the Bohmian trajectories look like in this case and in examples of "quantum tunneling" and scattering can be found in Chapter 7 of Albert (1992) and in Norsen (2013), respectively. As the diagram of trajectories in the two-slit experiment shows, every clearly stated question about how the particles come to have the locations they do has a clear and unique answer in this theory.

Earlier in this chapter, we made a sharp distinction between the configuration space of a set of N distinguishable particles in a three-dimensional space and that of a set of N indistinguishable particles. The first is represented by R^{3N}, and the second by ${}^{N}R^{3}$. We can now understand what ramifications this has.

If the particles are indistinguishable and we represent the quantum state by a field on ${}^{N}R^{3}$, then every set of N points taken from the space will evolve in a deterministic way encapsulated by the guidance equation. But if for mathematical convenience and simplicity, one prefers to work with a function on R^{3N}, that can be arranged. One just has to make sure that the evolution of the system with "permuted" particles is always the permutation of the original evolution. Since nothing yet prevents the wavefunction from having a completely different gradient in one configuration than it has in the permuted configuration, not every mathematically possible wavefunction is permissible as a representative of a quantum state. And the precise restriction that makes everything work out consistently is that the wavefunction on R^{3N} must be either symmetric or antisymmetric under exchange of indistinguishable particles. So this constraint on the wavefunctions of indistinguishable particles, which is simply postulated in standard quantum mechanics, falls out as a theorem in Bohmian mechanics.

Bohmian mechanics recovers all the predictions of the quantum recipe, including violations of Bell's inequality. It therefore must have some nonlocality built in. The nonlocality does not manifest itself in any single-particle experiment, for in such a case, the quantum state can be regarded as a localized field on space that locally guides the particle trajectory. It is only when multiple particles are involved and the quantum state cannot be understood as a local field on space that the nonlocality of the theory becomes manifest. To better understand this, we need to think more clearly about how the theory applies to subsystems of larger systems.

The Universal Quantum State and the Wavefunction of a Subsystem

Ultimately, the entire physical universe is one large interacting entity. But as a practical matter, we can never treat it as such. In everyday life and in laboratory practice, we treat subsystems of the universe as essentially independent of the rest of the universe. Since this is clearly an effective way of proceeding, physics should explain its effectiveness. How does it do so?

In a physical theory that postulates only local beables with local dynamics, the situation is clear. In such a theory, one can specify a subsystem of the universe simply by indicating a region of space (or space-time). Since all beables of the theory are local, the entire physical state of the region is determined by the local beables in the region. And if the dynamics only allows for "local action" (meaning that changes in the physical state are governed by differential equations in space and time), then any influence on the system from outside it must come through one of the boundaries. By isolating the system from the environment and not letting anything across the boundary, one gets a closed system that obeys the fundamental physical laws. If such complete isolation is not possible, still the outside influences can be accounted for by modeling effects across the boundary.

In a letter to Max Born, Einstein forcefully articulated a vision of physics committed to accepting only local beables and only local physical laws:

> If one asks what, irrespective of quantum mechanics, is characteristic of the world of ideas of physics, one is first of all struck by the following: the concepts of physics relate to a real outside world, that is, ideas are established relating to things such as bodies, fields, etc., which claim "real existence" that is independent of the perceiving subject—ideas which, on the other hand, have been brought into as secure a relationship as possible with the sense data. It is further characteristic of these physical objects that they are thought of as arranged in a space-time continuum. An essential aspect of this arrangement of things in physics is that they lay claim, at a certain time, to an existence independent of one another, provided these objects "are situated in different parts of space." . . . This principle has been carried to extremes in the field theory by localizing the elementary objects on which it is based and which exist independently of each other, as well as the elementary laws which have been postulated for it, in the infinitely small (four-dimensional) elements of space.[3]

Einstein's comment about the laws themselves being postulated for the small patches of space-time refers to the dynamics of classical field theory being local in the sense mentioned above.

This way of treating subsystems fails if irreducibly nonlocal beables exist or the laws are not local. If there are both local and nonlocal fundamental beables, then one can still indicate a region of space-time, but it is no longer clear what is meant by "the physical state of the system" in that region. And if the laws allow

[3]Einstein quoted in Born (1971), pp. 170–71.

for "action at a distance," so that events in one region can influence the physics in distant regions without a continuous chain of physical changes between them, then there will be no autonomous physical dynamics of the subsystem. Einstein thought that physics simply could not be done in such a situation, but as we will see, quantum mechanics itself proves him wrong.

Since we have some local beables—the particles—in the pilot wave picture, we at least have somewhere to begin. If we try to indicate a subsystem of the universe by indicating a space-time region, there is a definite fact about which particles are in that region. Or, even more directly, we might indicate a subsystem by picking out a particular set of particles, wherever they might happen to be. The "state of the system" will be clear to this extent: The subset of particles will have its own determinate configuration, fixed by where the particles are.

But to employ quantum theory, one needs more than that. One also needs a quantum state for the subsystem. And while it is easy to specify what the local beables of the subsystem are—given the totality of local beables and a characterization of the subsystem—it is not at all clear what the quantum state or the wavefunction of the subsystem should be, *even given the wavefunction or quantum state of the entire universe*. The complete collection of local beables divides simply into various "parts"—the subsets—but the universal quantum state has no spatial parts (or, indeed, any obvious division into parts at all).

One case where it does is when it is a product state. Suppose there are 100 particles, so the quantum state is defined on the configuration space $(q_1, q_2, \ldots, q_{100})$, where the qs are the positions of the indicated particles.[4] It might happen that the universal wavefunction $\psi(q_1, q_2, \ldots, q_{100})$ can at a certain time be factored into the product of two wavefunctions:

$$\psi(q_1, q_2, \ldots, q_{100}) = \phi(q_1, q_2, \ldots, q_{25})\, \xi(q_{26}, \ldots, q_{100}),$$

[4]Each individual q would be represented by a trio of real numbers, e.g., x_{q1}, y_{q1}, z_{q1}.

for example. In such a case, one could attribute the wavefunction $\phi(q_1, q_2, \ldots, q_{25})$ to the subsystem consisting of the first 25 particles and the wavefunction $\xi(q_{26}, \ldots, q_{100})$ to the subsystem consisting of the rest. And so long as the total wavefunction remains in an unentangled product state of this sort between the subsystems, each subsystem will obey exactly the laws of the theory relative to its own wavefunction. That is, the particle configuration of the first subsystem will evolve in accord with the guidance equation applied to $\phi(q_1, q_2, \ldots, q_{25})$, and the particle configuration of the second subsystem will evolve in accord with the guidance equation applied to $\xi(q_{26}, \ldots, q_{100})$.

Although this is interesting, it is of very limited value. Since the wavefunction in this theory never collapses, systems get more and more entangled. Even if two systems start out in a product state, interactions can easily entangle their wavefunctions. What then?

The pilot wave theory (but not theories that lack particles with definite locations in their ontologies) has further resources here. Let's review both what we have and what we want.

At a fundamental physical level, what we have is a universal quantum state represented by the wavefunction $\psi(q_1, q_2, \ldots, q_{100})$ and a set of 100 particles, each of which always has a precise actual location in space. The q_i used in the wavefunction are variables that take particular locations as values. We will represent the actual locations of the particles with capital letters: $Q_1, Q_2, \ldots, Q_{100}$. This, at any given time, is what we have to work with. And what we want is some sort of wavefunction defined just for the subsystem comprising (say) the first 25 particles. So abstractly, what we want is a scalar or spinor function of the form $\phi(q_1, q_2, \ldots, q_{25})$. The question is: Given what we have to work with, is there any obvious way to get what we want?

Indeed there is! Since the particles all have actual locations at any given time, the obvious thing to do to get rid of a variable in the universal wavefunction is to plug in the *actual location of the corresponding particle*. That is, from what we have to work with we can define the *conditional wavefunction* of our subsystem as follows:

$$\phi_{\text{cond}}(q_1, q_2, \ldots, q_{25}) =_{\text{df}} \psi(q_1, q_2, \ldots, q_{25}, Q_{26}, Q_{27}, \ldots, Q_{100}).$$

The conditional wavefunction is a well-defined mathematical object constructed from the mathematical representation of the universal quantum state (i.e., the universal wavefunction) and the mathematical representation of the actual positions of the particles not in the subsystem of interest. We can call these other particles "the environment" of the subsystem.

From an ontological point of view, then, the conditional wavefunction does not postulate anything new. The fundamental ontology of the theory still is completely specified by just the universal quantum state, the space-time structure, and the particles (which always have actual locations in space-time). In one sense of that multiply ambiguous term, the conditional wavefunction "emerges" from the fundamental ontology. The interesting question for us now is exactly how the conditional wavefunction, as a mathematical object, *behaves*.

The time evolution of the conditional wavefunction is a complicated business. That is because the ingredients from which it is constructed—the universal wavefunction and the locations of particles in the environment—each have their own time evolution. The universal wavefunction is always governed by the Schrödinger equation: It never collapses. And the particles in the environment always move in accordance with the guidance equation (which governs the complete configuration of particles). So the fundamental dynamical equations of the theory determine (given a particular universal quantum state and a particular universal configuration) what the conditional wavefunction does. Let's consider several contrasting cases.

First, consider the Double Slit with Monitoring, assuming that the initial universal wavefunction (in this case, just the wavefunction of the electron and the proton) is a product state. As we have seen, this product state becomes an entangled state via the interaction potential, and the universal wavefunction after the electron has passed through the slits is the entangled state

$$\frac{1}{\sqrt{2}}\phi_{upper}(x_e, y_e, z_e)\,\xi_{up}(x_p, y_p, z_p) + \frac{1}{\sqrt{2}}\phi_{lower}(x_e, y_e, z_e)\,\xi_{down}(x_p, y_p, z_p).$$

But in Bohmian mechanics (unlike, e.g., GRW theory) the proton at that time is either actually in the upper part of the cavity or actually in the lower part. If it happens to be in the upper part, then the conditional wavefunction of the electron, which we get by plugging the actual position of the proton into this universal wavefunction, is just $\phi_{upper}(x_e, y_e, z_e)$, and if the proton happens to be in the lower part of the cavity then the conditional wavefunction of the electron is just $\phi_{lower}(x_e, y_e, z_e)$. (Why is this? Since $\xi_{up}(x_p, y_p, z_p)$ represents a proton definitely in the upper part of the chamber, it is essentially zero for locations outside the upper part. So if we plug in a location in the lower part, this term becomes essentially zero, annihilating the first term of the universal wavefunction.)

For example, if the proton actually moves upward in our experiment, then the conditional wavefunction of the electron is essentially the wavefunction on the plane picked out by the proton position. Figure 26 illustrates that the conditional wavefunction for the electron in this case is just the single-slit wavefunction for the upper slit.

Even though the universal wavefunction never collapses, the conditional wavefunction of a subsystem can collapse when the subsystem becomes entangled with the environment in the right way. Indeed, this is precisely the sort of entanglement required if the particle positions in the environment are to provide reliable information about the particle positions in the subsystem (i.e., if the subsystem has been "measured" by interaction with the environment). This Bohmian collapse of the conditional wavefunction corresponds closely to the (vaguely defined) collapse of the wavefuncton in the quantum recipe, and also, in a different way, to the sharply defined GRW collapses. But unlike the GRW collapses, these will always occur in this experimental situation, even though there are only two particles involved (GRW collapses will normally not occur in usual time scales with so few particles). And also unlike the GRW collapses, the collapse of the conditional wavefunction is continuous in time rather than abrupt. It happens quickly but at a calculable rate, since both the universal wavefunction and the particle positions evolve

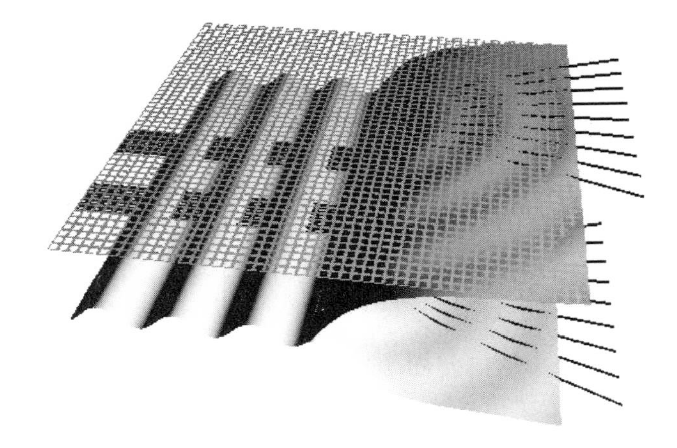

Figure 26

continuously, and these determine the conditional wavefunction. And unlike the GRW collapses, these collapses depend on entanglement of the system and environment, and so they only occur in "measurement-like" situations.

We have already seen that the GRW collapses occur where the nonlocality of that theory manifests itself. By virtue of a GRW collapse, the physical states of systems can change, and the changes in systems far from one another will be correlated. The collapse of the conditional wavefunction in Bohmian mechanics can similarly illustrate the inherent nonlocality of the theory.

Just looking first at the universal wavefunction, suppose we prepare a pair of particles A and B in a singlet state and let them separate. The spin part of the singlet state is $\frac{1}{\sqrt{2}}|z\uparrow>_A|z\downarrow>_B - \frac{1}{\sqrt{2}}|z\downarrow>_A|z\uparrow>_B$, and if we let A go far to the right and B far to the left, the entire wavefunction is

$$\frac{1}{\sqrt{2}}|z\uparrow>_A|\text{right}>_A|z\downarrow>_B|\text{left}>_B - \frac{1}{\sqrt{2}}|z\downarrow>_A|\text{right}>_A|z\uparrow>_B|\text{left}>_B.$$

By the standard quantum-mechanical recipe, one predicts from this wavefunction an equal chance for each particle to be deflected up or down by a z-oriented Stern-Gerlach magnet. Which way the deflection goes, in Bohmian mechanics, depends on exactly where the two particles actually are.

Suppose we send only particle A through such a magnet. By the usual Schrödinger evolution, the universal wavefunction is now

$$\frac{1}{\sqrt{2}}|z\uparrow>_A|\text{up-right}>_A|z\downarrow>_B|\text{left}>_B -$$
$$\frac{1}{\sqrt{2}}|z\downarrow>_A|\text{down-right}>_A|z\uparrow>_B|\text{left}>_B.$$

Again the usual quantum recipe applied to this state will yield equal 50-50 predictions if the z-spin of particle B is "measured," although the result for B is certain to be the opposite for A. But let's now calculate the *conditional* wavefunction for particle B, treating the distant particle A as part of the environment. Particle A has, in fact, actually been deflected up or actually been deflected down. If it has been deflected up, then the conditional wavefunction for B is $|z\downarrow>_B|\text{left}>_B$, and if it has been deflected down, then the conditional wavefunction for B is $|z\uparrow>_B|\text{left}>_B$. That is, the conditional wavefunction for B changes because of the experimental situation and location of the distant particle A, and the way it changes carries information about the sort of experiment that was carried out on A and about the outcome. If, for example, instead of a z-oriented magnet, A had been passed through an x-oriented magnet, then the conditional wavefunction for B would be an x-spin eigenstate rather than a z-spin eigenstate.

Since Bohmian mechanics is a deterministic theory, we have no problem calculating what would happen to a system in various different experimental circumstances. Sharply enough specified counterfactual or subjunctive claims have definite truth values. And in an EPR or Bell sort of experimental situation, those counterfactuals support behavior that anyone would call "action at a distance." For example, as we have just seen, whether the result of a particular experiment on the left comes out a certain way can depend on whether a certain experiment was carried out on the right. The manifest nonlocality of the theory in this sense—which before Bell's results could be considered grounds for objecting to the theory—explains how Bell's inequality will be violated for experiments done far apart in a world governed by this physics.

David Albert provides a striking example of how experimental outcomes in Bohmian mechanics can depend on distant experimental arrangements.[5] Consider again the spin version of the EPR experiment, where the two distant Stern-Gerlach magnets are perfectly aligned. Since Bohmian mechanics is a deterministic theory, the outcome of any particular run of the experiment—which will certainly give one "up" result and one "down" result—must be a consequence of the exact initial state of the particles together with the exact physical state of the experimental apparatus. But on which precise details of the physical situation does the outcome depend, and how?

One's first thought is that these EPR correlations can be explained simply. Which electron will be deflected up and which down is fixed in some direct way just by the initial state of the electrons. But Bohmian mechanics returns the predictions of standard quantum theory, including violations of Bell's inequality, so things can't be that easy. In a Stern-Gerlach experimental situation with a single particle, the outcome of the experiment is determined by the exact initial location of the particle. Suppose we prepare an x-spin up beam and pass it through a z-oriented magnet. Half the beam is deflected up and half down. Furthermore, in Bohmian mechanics, every single particle that is deflected up came into the device located above the midline of the magnet, and every single particle deflected down came in located below the midline. This follows from a simple fact about Bohmian mechanics: The trajectories in configuration space that describe how systems behave can never cross each other. Since the configuration of a single-particle system is given just by its position in space, that means that the possible trajectories of the particle through the Stern-Gerlach device cannot cross: The ones that eventually go upward must always be above all of those that eventually go downward.

[5] Albert (1992), pp. 155–60.

In the EPR situation there are two particles, and the configuration space for them is six-dimensional. If both the magnets are oriented in the z-direction, then half the time the right-hand particle will go up and the left down, and half the time the other way around. But which result actually occurs cannot be fixed in the same way merely by the antecedent locations of the particles. It is possible, for example, that both particles start out situated above the midline of their respective magnets. But it can't be that both particles are deflected upward. So what physically determines the outcome?

The answer, as Albert shows, is that the first particle that reaches its detector behaves just like the single particle discussed above: It exits the magnet moving upward if it enters above the midline and downward if it enters below. And the second particle displays the opposite behavior *no matter where it is located*. This corresponds to the fact that the conditional wavefunction of the second particle, after the first particle has passed through its device, will either be $|z\uparrow>$ or $|z\downarrow>$ (depending on the first outcome). So if in the initial state both particles start out above the midline, the exact experimental outcome will depend on which particle passes through its Stern-Gerlach magnet first. If the right-hand particle interacts first, the outcome will be up on the right and down on the left; if the left-hand particle interacts first, the result will be up on the left and down on the right.

We therefore can see how the outcome on one side depends, in a precise way, on the experimental situation far away. Given some initial conditions for the electrons, the electron on the right subject to a z-oriented magnet will be deflected up if no experiment is done on the left, deflected down if a z-oriented experiment is done on the left at an earlier time, deflected up if an x-oriented experiment is done on the left at an earlier time, and so forth. The nonlocality of the theory is manifest.

(We have just described how these experiments are treated in the nonrelativistic theory. In a nonrelativistic setting, the chance of the particles reaching the magnets at exactly the same time is essentially zero: One will always arrive before the other. But in a relativistic space-time, sufficiently distant experiments have no definite time order, and this sort of approach cannot be straightforwardly adopted. This raises profound issues. But at this point we just want to understand the nonrelativistic theory.)

So far we have seen how the notion of a conditional wavefunction allows us to ascribe a wavefunction to a subsystem of the universe, and how that wavefunction can, at times, collapse in just the way the quantum recipe requires. That leaves one last question. In actual experimental situations, we ascribe a wavefunction to a subsystem and use Schrödinger's equation to evolve the wavefunction until a "measurement" occurs. Under what conditions will the conditional wavefunction of a subsystem obey Schrödinger's equation (rather than suffering a collapse of some sort)?

In one case, the answer here is straightforward. If the quantum state of the universe is a product state $\phi(q_1, q_2, \ldots, q_{25})\, \xi(q_{26}, \ldots, q_{100})$, then the conditional wavefunction of the subsystem consisting of the first 25 particles is $\phi(q_1, q_2, \ldots, q_{25})$, and the conditional wavefunction of the rest of the universe is $\xi(q_{26}, \ldots, q_{100})$. If the universe starts out in such a product state and there are no interaction terms between the two systems in the Hamiltonian, then the universal wavefunction will always be a product state. Each conditional wavefunction will evolve by Schrödinger's equation, with no collapses. In this situation, the subsystems are unentangled with each other. Since any procedure we would call a "measurement" requires an entangling interaction between systems, we would say that the "environment" has not "measured" the system, and the conditional wavefunction of the system does not collapse.

But this straightforward case is too specialized to be of much use. Given the pervasive interactions among systems through the history of the universe, it is doubtful that the universal wavefunction is any sort of product state. Fortunately, a weaker condition also guarantees that the conditional wavefunction of a system will not suffer any collapse. Suppose that the universal wavefunction $\psi(q_1, q_2, \ldots, q_{100})$ is entangled but can be written as the superposition of a product state with some other state:

$$\psi(q_1, q_2, \ldots, q_{100}) = \phi(q_1, q_2, \ldots, q_{25})\, \xi(q_{26}, \ldots, q_{100}) + \psi_{\perp}(q_1, q_2, \ldots, q_{100}),$$

where $\psi_{\perp}(q_1, q_2, \ldots, q_{100})$ is "nonoverlapping" to $\phi(q_1, q_2, \ldots, q_{25})\, \xi(q_{26}, \ldots, q_{100})$ in the following sense: There are no points in configuration space where both $\psi_{\perp}(q_1, q_2, \ldots, q_{100})$ and $\phi(q_1, q_2, \ldots, q_{25})\, \xi(q_{26}, \ldots, q_{100})$ are significantly greater than zero. And further suppose that the actual configuration of the universe—where the particles actually are—corresponds to a region of configuration space where $\phi(q_1, q_2, \ldots, q_{25})\, \xi(q_{26}, \ldots, q_{100})$ is nonzero and $\psi_{\perp}(q_1, q_2, \ldots, q_{100})$ is (nearly) zero. Then (as you can prove!) the conditional wavefunction of the small subsystem is $\phi(q_1, q_2, \ldots, q_{25})$, and the conditional wavefunction of the environment is $\xi(q_{26}, \ldots, q_{100})$. And as long as the two subsystems do not interact, each conditional wavefunction will evolve without collapsing. So the appropriateness of using the Schrödinger equation for the subsystem is again justified.

There is still one lingering issue. By the linearity of Schrödinger's equation, we know that the time evolution of $\psi(q_1, q_2, \ldots, q_{100})$ can be recovered by adding the time evolution of $\phi(q_1, q_2, \ldots, q_{25})\, \xi(q_{26}, \ldots, q_{100})$ to the time evolution of $\psi_{\perp}(q_1, q_2, \ldots, q_{100})$. But just because $\psi_{\perp}(q_1, q_2, \ldots, q_{100})$ is nonoverlapping with $\phi(q_1, q_2, \ldots, q_{25})\, \xi(q_{26}, \ldots, q_{100})$ at one moment of time, it does not follow that at all future times the time evolution of $\psi_{\perp}(q_1, q_2, \ldots, q_{100})$ will be nonoverlapping with the time evolution of $\phi(q_1, q_2, \ldots, q_{25})\, \xi(q_{26}, \ldots, q_{100})$. And if at a later time, these two parts of the wavefunction come to overlap in configuration space, then our argument does not apply.

Let's get a concrete sense of exactly what this means. Look back again at Figure 25. On account of entanglement between the electron and proton, the wavefunction divides into two pieces that are effectively nonzero only in very different regions of configuration space. One piece is the branch going up, and the other is the branch going down in the figure. But on any particular run of the experiment, the actual configuration of the electron and proton follows a unique trajectory through configuration space. The lines represent different possible trajectories. Which trajectory occurs on a given run depends on the exact initial location of the particles.

Now suppose the actual trajectory in a given run is one of the upper lines: The electron has gone through the upper slit, and the proton has moved to the top of its chamber. The lower branch of the universal wavefunction still exists, since the universal wavefunction never collapses. But the way that the electron and proton continue to move depends only on the local gradient of the wavefunction, that is, the gradient *where the actual configuration is*. So the lower branch of the wavefunction, being far away in configuration space from that actual configuration, has no immediate influence on how the actual configuration will change. For all practical purposes, we can ignore the lower branch altogether.

But the lower branch still exists whether or not we attend to it, and it continues to evolve via Schrödinger's equation. And it is possible that this future evolution brings that branch back to the region of configuration space where the configuration actually is. If that happens, we can no longer just ignore the existence of that branch, because it will have observable effects on the behavior of our system.

The Mach-Zehnder interferometer is a real physical example of exactly this sort of reunion of separated branches of the wavefunction and its observable physical effects. Since the interferometer experiment involves only one particle, the configuration space for the system is isomorphic to physical space, and we can (taking due care not to be misled) picture the wavefunction of the system as a field in physical space. The wavefunction always propagates along both paths through the interferometer. We know that something like this is physically necessary, since the behavior of every single electron on every single run is sensitive to the physical conditions along both paths. But the electron itself, in the Bohmian theory, always travels along one path or the other. And while it is on one path, its behavior is unaffected by the branch of the wavefunction associated with the other path. If, for example, the electron is actually on the lower path, then we can calculate its trajectory along that path while completely ignoring the existence of the other part of the wavefunction.

But once the two paths are brought back together in space, both branches of the wavefunction play a role in determining how the electron behaves. Once the two branches come together in configuration space again, they will superpose and create constructive or destructive interference. That interference is essential to the prediction of the phenomena.

So it is a decent rule of thumb that once a wavefunction can be expressed as a superposition of wavefunctions that are effectively nonzero only in nonoverlapping regions of configuration space, one can just ignore any parts of it that are distant in configuration space from the actual configuration. But it is essential that this rule of thumb not be regarded as without exceptions. Many of our characteristically quantum-mechanical phenomena depend on the later recombination of such parts of the wavefunction and the resulting interference.

Under what conditions is such later interference physically difficult to achieve? The more entanglement there is, the harder it is to get branches of the wavefunction back together in configuration space. If a system contains 1,000 entangled particles, two branches of the wavefunction will be far away from each other in configuration space if even a single particle has a very different location in each branch. Our intuitions about the likelihood of accidental later interference here—the chance of a branch of the wavefunction that we had written off as physically irrelevant influencing the later evolution of the actual configuration—can be led astray by our unfamiliarity with very-high-dimensional spaces. Waves in a low-dimensional space (e.g., a swimming pool) can easily be accidentally reflected or refracted so as to interfere. But waves in a 10^{23}-dimensional space, like the configuration space of a macroscopic body, have so many different directions to move in that they are highly unlikely to ever accidentally interfere once they have become separated. That is why quantum interference effects are only noticeable in systems with very few particles or very few relevant physical degrees of freedom. Once a lot of degrees of freedom are in play, later interference between currently distant branches is unlikely to happen without someone taking pains to create it. The two branches in Figure 25 will typically never run across each other (in configuration space) again.

Bohmian Mechanics and the Eight Experiments

As already stated, one really understands a proposed physical theory of quantum-mechanical phenomena only if one understands how it yields the observed outcomes in our eight experiments. We have already covered most of them, but a quick review is in order.

Experiments involving only one particle as the target system are simple. The mathematical wavefunction represents the quantum state of the system, which always evolves linearly and never collapses. The particle trajectory is determined by its initial position and the gradient of the wavefunction via the guidance equation. The crests and troughs in our representation of the wavefunction indicate points with equal phase, and the gradient is directed orthogonal to these crests and troughs. The wavefunction evolves rather like a water wave, with the usual diffraction and interference, and the trajectories for the Single Slit and Double Slit experiments are shown in Figure 24. The so-called wave/particle duality is resolved in a trivial way: Both a wave and a particle are present. The locations of the small marks or flashes are determined by the locations of the impinging particles, and the wavelike interference and diffraction is due to the wavelike evolution of the quantum state.

We have discussed at length the effect of adding the monitoring proton and illustrated possible trajectories in configuration space in Figure 26. The interference bands disappear because of the separation of the branches of the wavefunction in configuration space. When spinors are added, entanglement of the spin and spatial degrees of freedom yield the most obvious trajectories through a Stern-Gerlach magnet: Particles that exit via the upper path came through the upper part, and particles that exit through the lower path come through the lower part. Since the wavefunction never collapses, it is still possible for parts of the quantum state associated with the road not taken to recombine and later influence the system. This possibility accounts for the interferometer results.

The EPR experiment already fully displays the nonlocality of the theory. Since the observed outcome on one side can depend on the experimental arrangements on the other—no matter how far away it is—it is not surprising that Bell's inequality can be violated. To derive the predictions, it is enough to see that if a pair of electrons is prepared in the singlet state and one is subjected to a Stern-Gerlach magnet oriented in some direction, the conditional wavefunction of the other will become an eigenstate of spin in the direction of that distant magnet. The second particle's motion is determined by this conditional wavefunction in accordance with the guidance equation. The desired correlations between the sides follow, even when the magnets on the two sides are not aligned.

As compelling as these observations are, there are further gaps in the account to be filled in. We have argued that the pattern of arrival of electrons at the screen in a Double Slit experiment, for example, will be just the interference pattern predicted by the quantum recipe. But of course, we never directly observe the arrival of any individual electron anywhere. To get an observable outcome, the electron must somehow trigger a magnifying cascade of events in its vicinity, leading to a macroscopic change in the apparatus. A mark on film must be made, or a pointer must move one way or another, for example. And to explain how this happens in a principled way, the apparatus itself must be treated as a physical system governed by the same laws as the electron. This is what it takes to solve the measurement problem.

The prospect of completing this account may seem dim. The experimental equipment contains a tremendous multitude of elementary particles. No one could ever know, write down, or calculate with the exact wavefunction of any such system. So how can the measurement problem really be solved?

Fortunately, the general character of the fundamental physical laws—in particular, the linearity of Schrödinger's equation—comes to the rescue. It is true that we can't write down the complete wavefunction of the Stern-Gerlach magnet-plus-screen that constitutes a spin-measuring apparatus. But we can test a given apparatus to see whether it always reacts in one way when an electron prepared as z-spin up is fed into it and reacts in an observably different way when a z-spin down electron is fed in. If so, then we know that the wavefunction evolutions must be

$$|z\uparrow>_e|\text{ready}>_d \Rightarrow |\text{indicates up}>_{e+d} \text{ and}$$

$$|z\downarrow>_e|\text{ready}>_d \Rightarrow |\text{indicates down}>_{e+d},$$

where the arrow indicates time evolution and the final state is a state of the electron-plus-device system. The two final states $|\text{indicates up}>_{e+d}$ and $|\text{indicates down}>_{e+d}$ must be confined to widely separated, nonoverlapping regions of configuration space, since the two final configurations of the apparatus differ in some macroscopic way.

Given these features of the apparatus and the linearity of Schrödinger's equation, we know how the initial wavefunction $|x\uparrow>_e|\text{ready}>_d$ must evolve (i.e., what happens to the wavefunction when we feed an x-spin up electron into a z-oriented Stern-Gerlach device). Since $|x\uparrow>_e$ is the superposition $z\uparrow>_e + \frac{1}{\sqrt{2}} z\downarrow>_e$, the final wavefunction must be

$$\frac{1}{\sqrt{2}}|\text{indicates up}>_{e+d} + \frac{1}{\sqrt{2}}|\text{indicates down}>_{e+d}.$$

And since the configurations associated with the two outcomes differ with respect to the positions of many particles in the apparatus, the wavefunction ends up with two equal-sized lumps in two widely separated regions of configuration space.

Recall the feature of the Bohmian trajectories mentioned above: They follow the probability current defined by the evolution of the wavefunction. Since half of the squared amplitude of the initial wavefunction evolves to $|\text{indicates up}>_{e+d}$ and half to $|\text{indicates down}>_{e+d}$, it follows that according to this measure, half of the Bohmian trajectories end up in the $|\text{indicates up}>_{e+d}$ region of configuration space and half in the $|\text{indicates down}>_{e+d}$ region. But then by this measure, half of the possible trajectories associated with this experimental situation yield an apparatus in the "indicates up" configuration and half yield an "indicates down" configuration. This sounds very close to the content of the prediction of the quantum recipe.

But if you think carefully about it, that's not quite enough for what we want. After all, in any single run of our experiment, we never get a result that corresponds to 50% of anything. All we ever get on a single run is a single result: an "indicates up" configuration of the apparatus or an "indicates down" configuration. What we really need to explain—what counts as the successful empirical content of the quantum recipe—necessarily encompasses more than any single such experiment. What we really need to account for are the observed *empirical frequencies*: the fact that if we repeat the experiment many times, about half the time we get an "indicates up" outcome and about half the time an "indicates down" outcome.

How does the theory account for these empirical frequencies? We have seen that if we use the squared amplitude of the wavefunction as a measure over configuration space at the start of our experiment, half of the possible Bohmian trajectories yield an apparatus in an "indicates up" state and half yield an apparatus in an "indicates down" state. But suppose instead we ask: If we repeat this kind of experiment 1,000,000 times and measure the initial configuration space for this compound experiment using the squared amplitude of the initial wavefunction, what measure of the possible initial conditions yield a set of results in which almost 50% of the outcomes are "indicates up" and almost 50% are "indicates down"? To make things exact, what is the measure of initial configurations that yield between 49.6% and 50.4% "indicates up" results? Any such result (unlike a result from a single experiment) would be taken to constitute evidence for the accuracy of the quantum recipe. How much of the initial configuration space of the system—the possible initial exact particle locations—leads statistics in this range?

If we measure the configuration space by the squared amplitude, more than 99.999999999% of the configuration space leads to statistics in this range. That is, as measured by the squared amplitude, overwhelmingly most of the possible initial configurations yield the right empirical frequencies, where by "right" we mean between 49.6% and 50.4% "indicates up" results. In this precise sense, empirical statistics in this range are *typical* for such a system, where the "measure of typicality" used is the squared amplitude of the universal wavefunction.

It is critical in this argument that "typical" here means that the measure of possible initial states leading to this sort of result is very close to 1. Because the number .99999999999 is so close to 1, we can see that the outcome remains typical in this sense, even if we switch to another measure over the configuration space, so long as the new measure is not wildly different from the old one. Any vaguely similar measure would give the same result: The right empirical statistics are typical. For example, since it can be proven that Born statistics are typical relative to the Ψ-squared measure of the initial configuration space, it is also true for "overwhelmingly most" possible initial configurations using $|\Psi|^4$ as the measure, or $|\Psi|^6$, or $|\Psi|^8$.

We do not get the result that Born statistics—the statistics given by $|\psi|^2$, where ψ is the conditional wavefunction of a subsystem—are typical, because we use $|\Psi|^2$ (where Ψ is the initial universal wavefunction) as the measure of typicality. If using $|\Psi|^4$ as the measure of typicality yielded $|\psi|^4$ statistics as typical (an argument that Detlef Dürr calls "garbage in, garbage out"), then Born statistics could only be recovered by fine-tuning the measure of typicality. But that is not the situation at all. It is rather like this: Take the initial configuration space with an initial universal wavefunction and ask of each possible initial configuration whether is would yield Born statistics for subsystems. If it would, paint it green; if it wouldn't, paint it red. Now if by the $|\Psi|^2$ measure on the configuration space 99.999999999% of the points are green, and the whole configuration space looks essentially solid green, then it will still look solid green under all sorts of distortions of that measure. A wall that has microscopic red specks but looks solid green still looks solid green if you put on distorting glasses that change the apparent areas of the wall. It is only by looking through a specifically sited and focused high-power microscope that the wall might appear to be noticably red.

The squared-amplitude measure of configuration space has another feature that recommends it as a good way to make sense of what "overwhelmingly most" of the possible configurations of a system means. Suppose that, at a certain moment, we want to quantify what proportion of the possible configurations available to a system lie in a certain region of configuration space. Since there will be infinitely many possible configurations in the region, we need a measure to quantify the proportion. At time t_0, the measure ascribes some such proportion to S_0, the set of configurations in that region. For example, according to the measure, 75% of the possible configurations lie in that region. Now let all configurations in S_0 evolve in time by means of the guidance equation until t_1. The set S_0 will evolve into a new set of configurations S_1. And if we use the same rule to ascribe a measure to S_1 at t_1, it had better come out to be the very same size as S_0 was at t_0. For example, if 75% of the possible configurations available at t_0 were in S_0, then 75% of the configurations available at t_1 should be in S_1. For the set of systems in S_0 at t_0 is just the set of systems in S_1 at t_1. All systems originally in S_0 end up in S_1, and no new systems are added. Any measure that has this formal property—the measure of sets of systems remains constant as they evolve under the time evolution—is called *equivariant*.

The squared-amplitude measure of configuration space is an equivariant measure. That is also a consequence of the trajectories following the probability current defined by the evolution of the wavefunction. Equivariance is a necessary condition for any coherent measure of "how large" a set of systems—represented by a region in configuration space—is.

In sum, as judged by the equivariant squared-amplitude measure, overwhelmingly most of the possible initial configurations will evolve to display the empirical frequencies predicted by the quantum recipe.[6] The recipe predictions will, in this sense, be typical in Bohmian mechanics. If we accept this as accounting for the observed statistics, then Bohmian mechanics accounts for the phenomena predicted by standard nonrelativistic quantum theory. Furthermore, it does so in a way that provides a detailed physical account of what is going on at the microscopic scale in our experiments. Particles move around in accordance with the guidance equation, and the guidance equation employs the

[6] This brief discussion of is only the tip of a very complex mathematical discussion that would go far beyond the level of this book. Interested readers with strong mathematics can consult Dürr, Goldstein, and Zanghì (1992) or Dürr and Teufel (2009).

wavefunction with pure (noncollapse) linear Schrödinger evolution. The "collapse" of the conditional wavefunction follows from this fundamental physics by analysis. There is also an effective collapse of the wavefunction (we can safely permanently discard parts of it) due to the fragmentation of the universal wavefunction into separated pieces in configuration space, as in the Double Slit with Monitoring experiment. Using the theory, we can even draw specific conclusions for particular experiments. In the Double Slit experiment, for example, if a mark forms on the upper part of the screen, the electron went through the upper slit; if it forms on the lower part, the electron went through the lower slit (see again Figure 24). No similar claim is true in the GRW theory. In that setting, the electron no more "goes through" one slit rather than the other on any run, no matter what the outcome.

There is no problem accounting for measurements and their outcomes. Measurements are just interactions between one physical system and another, governed by the same universal laws. Sometimes, by virtue of the interaction, the configuration of one system will change in different ways, depending on the interaction. If this system is large enough, and the different possible configurations are distinct enough, we can tell by looking at which way the experiment came out. There is nothing magical about experimental apparatus or measuring devices: they are just physical systems like everything else. And if one asks what, if anything, a particular experiment measures, the answer is determined by pure physical analysis. If the observable outcome depends on some feature of the initial state of the interacting system, then the outcome provides information about that feature. What information it provides depends on the details of the interaction. Such questions are settled by using the theory to analyze the interaction.

So there is no measurement problem in Bohmian mechanics. Nor could there, in principle, be anything like a problem with Schrödinger's cat (or any other cat). Cats are made of particles, according to this theory, and the particles are always in some exact place moving in some exact way. An evolving configuration of many, many particles can unproblematically correspond to how we think a live cat is behaving, or how a dead cat is behaving (or to neither!). The supposedly problematic Schrödinger cat state is a state of the *wavefunction* or *quantum state* of the cat, a superposition of two macroscopically distinct states. According to Bohmian mechanics, the cat always has a quantum state, which never fundamentally collapses. But since the important role of the quantum state is to guide the motions of the particles, it doesn't matter at all that it doesn't collapse and that both branches of the wavefunction always exist. The branch that is far in configuration space from the actual configuration of particles in the cat becomes irrelevant for the cat's behavior.

In a nonrelativistic space-time, Bohmian mechanics provides an uncomplicated physics that accounts for all our experiments. There are particles that move around in accordance with a single, simple, deterministic law of motion. That law itself makes use of a quantum state of the system that always evolves by the familiar linear deterministic dynamics shared by all nonrelativistic quantum theories. The quantum state is a physically real, nonlocal entity in the theory; via its nonlocality, the motions of the particles get coordinated even when they are very far apart. That physical coordination yields violations of Bell's inequality. This is a nonlocal theory, and obviously so. But we know that we need some nonlocality if we are to recover what we take the phenomena to be: violations of Bell's inequality for outcomes of experiments performed at great distances from one another.

Since the pilot wave approach is simple, has no conceptual difficulties, and recovers the content of the quantum recipe in the nonrelativistic setting, one might wonder why it is not at least discussed in physics textbooks.[7] This question requires a sociological answer. Some historical research relevant to answering that question is provided in the Further Reading section at the end of this chapter. But to raise this question, one must first be convinced that the theory does work. We have seen the outlines of the theory, and much more mathematical and physical detail can be found in the literature listed in Further Reading.

[7] For a compelling expression of this question, see Bell's "On the Impossible Pilot Wave" in Bell (2004), Chapter 17.

There is still much technical work to be done on the pilot wave approach. It is not straightforward to adapt this sort of theory to a fully relativistic regime. We have already seen why: In the non-relativistic version, sometimes the outcome of an experiment can depend on the time order of distant experiments. But in a relativistic space-time, no time order at all exists between sufficiently separated regions. So a way forward requires either a different sort of implementation of the basic idea or a reconsideration of the relativistic account of space-time structure. But the success of the nonrelativistic theory provides a motivation to continue developing pilot wave accounts.

Further Reading

Introductions to Bohmian mechanics exist at all levels of technical and mathematical sophistication. The following have already been cited. Chapter 17 in Bell (2004) is quite advanced. A less technically demanding overview is chapter 20 in Bell (2004). A mathematically rigorous account is Dürr and Teufel (2009). Articles at different levels of mathematical sophistication are collected in Dürr, Goldstein, and Zanghì (2013).

Many physical and philosophical questions about Bohmian mechanics are discussed in Cushing, Fine, and Goldstein (1996) and in Bricmont (2016). Both the theory and historical questions are treated in Cushing (1994). Many of the sociological questions about the treatment of the theory are addressed in Beller (1999).

CHAPTER 6

Many Worlds

WITH HIS CHARACTERISTIC succinctness, John Bell summed up possible reactions to the problem of Schrödinger's cat, as understood by Schrödinger:

> He [Schrödinger] thought that she [the cat] could not be both dead and alive. But the wavefunction showed no such commitment, superposing the possibilities. Either the wavefunction, as given by the Schrödinger equation, is not everything or it is not right.[1]

The possibility that the wavefunction is not everything (i.e., that it is not informationally complete, so the wavefunction of a system does not determine all the physical features of the system) is implemented in Bohmian mechanics. The physical characteristics *not* reflected in the wavefunction in that theory (the particle locations) determine the macroscopic characteristics of the cat, including its state of health. Given only the history of particle positions through time, one can determine whether the cat lives or dies, even being ignorant of the wavefunction. Given only the history of the wavefunction through time, one cannot.

The possibility that the wavefunction as given by the Schrödinger equation is not right is embraced by objective collapse theories, such as GRW. The wavefunction in that theory is informationally complete: Knowing how the collapses occur suffices to determine the outcome of the experiment. But also in this case, the physics goes through the local beables. Since the distribution of matter density in the matter-density theory and the distribution of the flashes in the flash theory are determined by the behavior

[1] Bell (2004), p. 201.

of the quantum state, one can recover the distribution of local beables from the history of the wavefunction. The local beables then determine whether the cat lives or dies. In both these sorts of theories, there is an objective, physical, matter of fact about how the experiment comes out. At the end of the day, there is only one cat, which is either alive or dead.

Bell's categorization of solutions to Schrödinger's problem rests on acceptance of Schrödinger's assumption: The cat simply either lives or dies. Reject that assumption, and the problem as stated does not arise. In 1957, Hugh Everett inaugurated a new strategy for understanding the quantum formalism, which has come to be known as the Many Worlds interpretation. Our primary characterization of the general strategy can be drawn directly from Bell: In a Many Worlds theory, the wavefunction, as given by the Schrödinger equation, is both everything and right. What would such a physical world be like?

There are different ways to approach this question. Everett used interpretive principles that lead quickly to a Many Worlds picture, but the principles are difficult to defend. Let's rehearse what we know. Consider an experimental situation containing a *z*-oriented Stern-Gerlach apparatus coupled to a detector and a pointer so constructed that if a flash occurs on the upper part of the screen, the pointer moves to point up, and if a flash occurs on the lower part, the pointer moves to point down. If the entire physical state of the experiment is to be encoded in a wavefunction, then there must be at least one wavefunction that describes the state of the apparatus at the start of an experiment. Call this wavefunction $|\text{ready}>_a$. If we feed a $|z\uparrow>$ electron into the apparatus, then the initial wavefunction of the whole system is $|z\uparrow>_e|\text{ready}>_a$. And if the apparatus works as advertised, always ending up in this case with the pointer pointing up, then the Schrödinger evolution must produce a state $|\text{points up}>_{e+a}$, in which the pointer points up. Since we assume that the wavefunction is informationally complete, the final disposition of the pointer must follow from this state.

Similarly, if we feed in a $|z\downarrow>$ electron, then $|z\downarrow>_e|\text{ready}>_a$. must evolve by Schrödinger evolution into $|\text{points down}>_{e+a}$, where the latter is a state in which the pointer points down. What if we feed in an $|x\uparrow>$ electron? The answer is already determined. The initial state of the system is $|x\uparrow>_e|\text{ready}>_a$. But since $|x\uparrow>$ is a superposition of $|z\uparrow>$ and $|z\downarrow>$, we can with equal mathematical accuracy represent the initial state as $(\frac{1}{\sqrt{2}}|z\uparrow>_e + \frac{1}{\sqrt{2}}|z\downarrow>_e)|\text{ready}>_a$ or, equivalently, $\frac{1}{\sqrt{2}}|z\uparrow>_e|\text{ready}>_a + \frac{1}{\sqrt{2}}|z\downarrow>_e|\text{ready}>_a$. The linearity of the Schrödinger evolution implies that this state will evolve into $\frac{1}{\sqrt{2}}|\text{points up}>_{e+a} + \frac{1}{\sqrt{2}}|\text{points down}>_{e+a}$. Just as Bell says, this state shows no commitment to the pointer pointing one way rather than another. It is a superposition of states in which, by hypothesis, it points in different directions. Everett deployed his new interpretive principle here: If the quantum state of a system is a superposition of two different quantum states, then *both superposed states really exist*. Since $|\text{points up}>_{e+a}$ is by hypothesis a state in which the pointer points up, and $|\text{points down}>_{e+a}$ a state in which the pointer points down, at the end of the experiment, there is both a pointer pointing up and a pointer pointing down. The world has split into two "branches."

Everett's criterion for the real existence of a branch is extremely liberal: If the wavefunction $|\Psi>$ of a system is a superposition $\alpha|\Phi> + \beta|\Xi>$, then the "branches" $|\Phi>$ and $|\Xi>$ exist, and in each branch, everything physically exists that would exist if that were the entire state of the system. By this criterion, if the wavefunction of an electron is $|x\uparrow>_e$, then a branch containing an electron in the state $|z\uparrow>_e$ exists and a branch containing an electron in the state $|z\downarrow>_e$ exists. Indeed, since $|x\uparrow>_e$ is a superposition of up and down spin states in any direction except the x-direction, there would have to be branches with each of the states. According to Everett's account, every possible quantum state of any system contains infinitely many branches, since it can be expressed as a superposition in infinitely many ways. If one calls each such branch a "world," then Everett's theory is committed to infinitely many real worlds at all times, no matter what the quantum state of the universe happens to be.

It is hard to make sense of this account of branching, and nowadays Many Worlds theorists do not try. They do not believe that just because the wavefunction can be written as a superposition of states, those superposed states correspond to "real" or "actual" things. That criterion for the reality of a branch is too liberal. In its stead, modern Everettians appeal to *decoherence*.

David Wallace's *The Emergent Multiverse* (2012) is a paradigmatic statement of modern Everettianism. Wallace makes exactly the point above about superposition not being sufficient for branching and invokes decoherence as the condition that creates a multiplicity of "worlds."[2] To say that a wavefunction decoheres into two (or more) other states is to say more than that it is a superposition of those states. It is to say, in addition, that the elements of the superposition do not—and will not in the future—interfere. The interference of two wavefunctions, in turn, has a clear mathematical meaning. Given the superposition $|\psi(x,t)> + |\phi(x,t)>$, the two component states interfere at time T if $||\psi(x,T)> + |\phi(x,T)>|^2 \neq ||\psi(x,T)>|^2 + ||\phi(x,T)>|^2$. Conversely, the two component states strictly decohere for position at T if $||\psi(x,T)> + |\phi(x,T)>|^2 = ||\psi(x,T)>|^2 + ||\phi(x,T)>|^2$. In short, a superposition of two states strictly decoheres at T if the absolute square of the sum is the sum of the absolute squares. Most superpositions do not satisfy this requirement.

Consider, for example, the wavefunction at the screen in the Double Slit experiment. It is the superposition of two other wavefunctions: the wavefunction that would have existed if only the upper slit had been open and that which would have existed if only the lower had been open. But the absolute square of this superposition is not the sum of the components absolute squares on account of constructive and destructive interference.

Given this definition of decoherence at a given time T, we can define decoherence in general: A wavefunction has decohered into two branches at a time T_0 if it is a superposition of two wavefunctions $|\psi(x,t)>$ and $|\phi(x,t)>$ such that for all $t > T_0$, $||\psi(x,T)> + |\phi(x,T)>|^2 \approx ||\psi(x,T)>|^2 + ||\phi(x,T)>|^2$.

While Everett's condition for branching was too liberal, the decoherence condition is quite restrictive. Indeed, no superpositions satisfy this requirement if we demand strict decoherence

[2] Wallace (2012), p. 62.

at all future times. So modern Everettians do not require strict decoherence for branching; they require only approximate decoherence, as the $\approx$ in the definition testifies. The reason that no superpositions exactly satisfy the requirement is akin to the problem of tails in GRW: Wavefunctions tend to spread to have nonzero values at all points in configuration space, even if in most places those values are extraordinarily close to zero. But if two wavefunctions are both nonzero at some point in configuration space, they also interfere, so the squared amplitude of the sum is not the sum of the squared amplitudes. If decoherence is to do any work, only approximate decoherence can be demanded.

The requirement that the elements of a superposition not interfere at a specific time or in the future makes the condition even harder to satisfy. Consider Experiment 6, the interferometer. Halfway through the experiment, the wavefunction of the electron is $\frac{1}{\sqrt{2}}|z\uparrow>|\text{upper}> + \frac{1}{\sqrt{2}}|z\downarrow>|\text{lower}>$, with $|\text{upper}>$ being a wavefunction confined to the upper path and $|\text{lower}>$ a wavefunction confined to the lower path. Since $|\text{upper}>$ and $|\text{lower}>$ do not overlap (to any non-negligable degree) in configuration space, $|\frac{1}{\sqrt{2}}|z\uparrow>|\text{upper}> + \frac{1}{\sqrt{2}}|z\downarrow>|\text{lower}>|^2 \approx |\frac{1}{\sqrt{2}}|z\uparrow>|\text{upper}>|^2 + |\frac{1}{\sqrt{2}}|z\downarrow>|\text{lower}>|^2$: The wavefunction approximately decoheres into those components at that time. But this does not persist. When the two beams are brought back together at the end of the experiment, there is interference again. The existence of both components is essential to the explanation of the observable outcome of the experiment, and in this sense the two branches do not evolve independently of each other.

To address this issue, one refers not merely to decohering states but to decohering *histories*. A history specifies the state of a system at multiple times, and a set of histories decoheres if there are no appreciable interference terms between the histories. A history specifying the electron as on the lower path and then later as on the recombined path does not decohere from a history specifying the electron on the upper path and then on the recombined path exactly because of interference between the two later states on the recombined path. But on what basis could one be confident that two states not only do not interfere at a given time, but also will not come to interfere in the future?

The key observation solving this problem is the delicacy of the experimental arrangement needed to guide the two branches back to the same region of configuration space. If instead of a single electron whose wavefunction is a superposition of a part confined to the upper path and a part confined to the lower path it were a cat composed of 10^{24} particles, *every single one of the particles* would have to be meticulously guided to end up in the same spatial location no matter which path was taken. If even a single particle of those 10^{24} goes astray—ending up in one location via the lower path and another via the upper—then we have the Double Slit with Monitoring: the interference effects disappear. And if the cat is constantly interacting with its environment (as it will), the number of entangled particles to keep track of grows. For all practical purposes, states like these will permanently decohere.[3]

Demanding decoherence of branches solves the problem of too many branches, but perhaps at the price of too few. Since exact decoherence is not to be had, all one can ask for is approximate decoherence, with little interference between branches. But "little" is a vague term, so on this account, the branching structure is also vague rather than exact.

The claims made here about (approximate) decoherence are neither controversial nor unique to Many Worlds approaches. As defined, the universal wavefunction under a particular linear dynamics either approximately decoheres or it doesn't. The same approximate decoherence of the wavefunction occurs, for example, in Bohmian mechanics. But in Bohmian mechanics, this does not lead to multiple worlds in the sense of multiple cats at the end of Schrödinger's experiment. Bohmian cats are composed of particles, and the particles never split or branch or multiply: they simply follow one set of trajectories or another. The particle configuration at the end is either that of a live cat or of a dead cat. The characteristic flavor of the Many Worlds approach lies in

[3] Further discussion of why the dynamics of the wavefunction yields this sort of decoherence can be found in Wallace (2012), Chapter 3.

its commitment to the informational completeness of the wavefunction. If the wavefunction is informationally complete and it splits or branches, then the physical world—including the individual cat— must somehow also split or branch into a live version and a dead version. It is this astonishing claim that the Everettians embrace.

The Problem of Probability

We have the skeleton of a physical theory on the table. The theory is committed to a real quantum state, to universal linear evolution of the quantum state, and to the informational completeness of the wavefunction that represents the quantum state. These characteristics do not yet yield a complete theory, since the question of other ontology—in particular, local ontology in space-time—has not been addressed. We have seen how a given dynamics for the quantum state can be supplemented with different local beables in our discussion of the GRW theory. In each version of that theory, the wavefunction is informationally complete, but the versions still differ over the local ontology. The same issue arises for Many Worlds, but we will put that question off for the moment.

One main difference between GRW and Many Worlds concerns the understanding of probability in the theory. In GRW, the probabilities are easy to explicate: They are the probabilities for the quantum state to collapse in one particular way or another. The probabilities arise from the dynamics in a straightforward manner, since the dynamics itself is probabilistic. But no such understanding of probability talk is on offer for Many Worlds, since the dynamics of the quantum state is deterministic.

Bohmian mechanics also has a deterministic evolution of the quantum state, so perhaps it is the place to look for guidance about probabilities. But the probabilistic statements in Bohmian mechanics do not concern the quantum state at all. They concern the actual particle trajectories. The story here is more subtle. Given an initial quantum state, there is a set of possible initial conditions for the particles. Since the complete dynamics of the theory is deterministic, each possible initial condition implies particular outcomes for future experiments, including the statistics of collections of experiments. One then shows that for almost all initial conditions (by the relevant measure), these statistics will match the predictions of the quantum recipe. For this approach to make sense, the wavefunction cannot be complete (so a nontrivial set of possible initial conditions exists, even given a wavefunction) and particular initial states have to imply determinate statistics for outcomes. Since Many Worlds asserts that the wavefunction is complete, it cannot take this route.

How, then, are we to make sense of probability in the Many Worlds approach? Even granting a branching structure to physical reality, how is that structure related to any notion of *likelihood*? A concrete situation will bring the problem into focus.

Let's consider a modification of Schrödinger's experiment. In the new experiment there are two cages, each with its own diabolical device containing, at the moment, no cat. A *z*-oriented Stern-Gerlach apparatus sits outside the boxes, and things are arranged so that an electron coming out with an upward trajectory will trigger the device in Box A, while an electron coming out with a downward trajectory will trigger Box B. We are going to feed into the Stern-Gerlach apparatus an electron whose spin state is $\sqrt{\frac{2}{3}}\,|z\uparrow>_e + \sqrt{\frac{1}{3}}\,|z\downarrow>_e$. Our usual linearity arguments show that this state will evolve into

$$\sqrt{\frac{2}{3}}\,|z\uparrow>_e|\text{Box A triggered and Box B untriggered}> +$$

$$\sqrt{\frac{1}{3}}\,|z\downarrow>_e|\text{Box A untriggered and Box B triggered}>.$$

Since the positions of so many particles differ in the triggered and untriggered states, the two components of this final superposition decohere and will remain decohered. There will be branching.

The Many Worlds theorist asserts that in one of the decohered branches, Box A has been triggered and Box B has not; in the other, Box B has been triggered and Box A has not. This outcome of the experiment will occur with certainty if the Many Worlds account is true and can be predicted to occur with certainty by the experimenter, who knows how the experiment was designed.

The standard quantum recipe agrees about the time development and decoherence of the wavefunction so long as no appeal to Born's Rule is made. Of course, the sort of experimental set-up described would be considered a position measurement (and also spin measurement) of the electron, with the state of the boxes indicating whether the electron was deflected up or down. But if we imagine denying that status to the experiment and hence blocking appeal to Born's Rule, the result is that the quantum recipe would make no empirical predictions at all. Empirical predictions derive from the invocation of that rule, allowing one to ascribe probabilities to each of various possible, mutually exclusive outcomes. In contrast, the Many Worlds theorist is committed to denying that there will be one of two mutually exclusive possible outcomes. Instead, there will, with absolute certainty, be the decoherent branching outcome described above. So it is not clear how Many Worlds can recover or vindicate Born's Rule as it was originally proposed (i.e., as a way to assign likelihoods to alternatives).

Probabilities are assigned to a set of possible outcomes. If the set is mutually exclusive and jointly exhaustive, each outcome is assigned a real number between zero and one (inclusive), and probabilities sum to one. But in the Many Worlds theory, only one outcome of the experiment described above is possible: the branching structure given by the linear evolution of the wavefunction. So the probabilistic structure is trivial.

Some Many Worlds theorists have advocated a fallback position. Even if it is physically certain what the outcome will be and furthermore, the experimenter (having accepted Many Worlds) knows what that certain outcome will be, nonetheless a rational agent will act just like someone who is uncertain about the outcome and will assign different probabilities to different possible outcomes. This strategy seeks to vindicate behaving as if there were some uncertainty about the future, quantified by various probabilities, even when no such uncertainty exists.

To implement this strategy, we need to introduce an agent with various possible courses of action in our scenario. Imagine that you are the agent, and you are forced to make a horrible choice: You must either put your cat Erwin into Box A or into Box B before the experiment is done. You love Erwin. You want Erwin to survive. What should you do?

From the perspective of GRW theory, the right choice is clear. The GRW dynamics assigns a 2/3 chance that the diabolical device in Box A will be triggered, a 1/3 chance that the device in Box B will be triggered (and no chance that both or neither will be triggered). So you should put Erwin in Box B and cross your fingers. In the event that Box B is triggered and Erwin dies, you can console yourself that your choice was rational, even though it did not work out as you wished, and you didn't get what you wanted. Similarly for Bohmian mechanics. You know that the exact initial state will determine which box gets triggered, and you know that you don't know what that initial state is. But more possible initial states lead to Box A being triggered than Box B being triggered. If you are rational, you put Erwin in Box B and cross your fingers.

Standard decision theory says that one should calculate the *expected utility* of each possible action and choose the act with the greatest expected utility. The expected utility is calculated by weighting the value, or utility, of each possible outcome by its probability and summing these terms. It is easy to see that putting Erwin in Box B has a higher expected utility than putting him in Box A if we value his survival.

How should the Many Worlds theorist think about this? As far as the boxes go, everything is already determined. The world is about to "split" into two decohering branches, and Erwin will split as well, with a successor on each branch. Your only choice is which box Erwin goes in at the beginning. If you put Erwin in Box A, the world will branch, with one branch containing an Erwin-successor in A and A triggered and B untriggered (so the Erwin-successor will be dead) and another branch with an Erwin-successor in A and A untriggered and B triggered (so the Erwin-successor will be alive). If you put Erwin in Box B, then the world will branch, with one branch containing an Erwin-successor in B and A triggered and B untriggered (so the Erwin-successor will be alive) and another branch with an Erwin-successor in B and

A untriggered and B triggered (so the Erwin-successor will be dead). Your choice is a choice between bringing about the first outcome or bringing about the second. Which should you prefer?

Since the description of the two outcomes given above is symmetric substituting "A" for "B," there might seem to be no grounds to prefer one choice over the other. But we have left out one physical fact: The squared amplitude of the branch in which Box A (or rather: the Box A-successor on that branch) is triggered is higher than the squared amplitude of the branch in which Box B is triggered. The ratio of the measures of the squared amplitudes distinguishes the branches from each other. The question is: Why, as an agent, should you care about this proportional difference in squared amplitudes? And if you do care, should you prefer that the higher squared-amplitude branch contain an Erwin-successor in the triggered box and the lower amplitude branch contain an Erwin-successor in the untriggered box or the other way around?

The Many Worlds theorist would like to prove that the rational choice to make in this situation is the same as the rational choice made by an agent in a GRW world or an agent in a Bohmian world. But in those scenarios, the rationality of the act is tied to the fact that there will be only one Erwin-successor at the end of the experiment, and one wants that unique Erwin-successor (which we can just call "Erwin" without equivocation) to be alive. Putting Erwin in Box B maximizes the chance of succeeding, but it does not guarantee success. Since there will certainly be two Erwin-successors in the Many Worlds scenario, one alive and one dead, this justification for a rational choice is not available.

There is a further puzzle. As we have seen, the only possibly relevant difference between putting Erwin in Box A and putting Erwin in Box B is the relative squared amplitude of the branch on which the Erwin-successor lives compared to the branch on which the Erwin-successor dies. But exactly because of decoherence, the comparative values of the squared amplitudes can make no difference at all to the intrinsic character of either branch. Once they have decohered, the branches evolve independently of the existence of the other branches. So the relative squared amplitudes of the two branches can make no practical difference to anything or anyone on either branch. One what grounds, then, should it make any difference to the agent?

The technical challenge for the Everettian is to articulate some compelling "rationality principles" that imply that a rational Everettian agent must treat the squared amplitude of future branches exactly as the non-Everettian using Born's Rule (assigning a probability to the unknown future outcome) would. That is, the squared amplitude must enter into the decision-making procedure just as a probability would, yielding the same set of decisions. The rational Everettian will, to that extent, *act as if* the squared amplitude were a probability.

Unfortunately, the general scheme—lay down principles of "rationality" together with the Many Worlds physics and derive rational decision-theoretic behavior—has been implemented in different ways using different rationality principles. Furthermore, the list of principles and the proofs tend to be complex. David Deutsch, the originator of the strategy, used rationality principles couched in terms of "measurements" (associating the measurements with Hermitian operators).[4] David Wallace eschews all mention of measurements or operators. Instead, the acts available to the agent are represented by different Hamiltonians, and the consequence of choosing an act is implemented by allowing that Hamiltonian to generate the future branching structure. Since Wallace's approach coheres better with the physics, we will use it as our test case. It can provide a sense of how the general strategy is supposed to work and what conceptual issues it faces. But of necessity, our discussion is sketchy.

Let's return to Erwin. Intuitively, faced with the choice between putting Erwin in Box A and putting him in Box B, the right thing to do is to put him in Box B, maximizing his chance of survival. (Of course, an Everettian could just deny that this is the right choice in an Everettian universe, and our failure to appreciate that stems from our unawareness that our universe is Everettian. But no Everettian argues this way. Oddly, the hypothesis that there is a vast, heretofore unrecognized branching

[4]See Deutsch (1999).

structure to the physical world is supposed to have no surprising practical consequences for everyday life.) Since the Many Worlds Erwin will certainly have a living successor and a dead successor no matter what choice is made, one can't maximize the chance of survival in the normal sense. But suppose we change the physical set-up in the following way.

Instead of having the single electron on the up path immediately trigger the boxes, we send the up beam through a Stern-Gerlach magnet oriented in the *x*-direction. The up output beam of this second magnet is directed at a green trigger that will set off the device in Box A, and the down output beam is directed at a distinct red trigger that will also set off the device in Box A. Again, you have a choice: put Erwin in Box A or put Erwin in Box B. How should you think about this?

The branching structure that will result from either choice is clear. Instead of branching into two macroscopically distinct decoherent worlds, there will now be three. In one, the green trigger has gone off in Box A; in another, the red trigger has gone off in Box A; and in the last, the trigger has gone off in Box B. But unlike our original experiment, the squared amplitudes of these three branches are the same. Where should you put Erwin?

The state of the universe at the end if you put Erwin in Box A will be

$$\sqrt{\frac{1}{3}}|z\uparrow>_e|\text{Box A red triggered, Box B untriggered, Erwin dead}>+$$

$$\sqrt{\frac{1}{3}}|z\uparrow>_e|\text{Box A green triggered, Box B untriggered, Erwin dead}>+$$

$$\sqrt{\frac{1}{3}}|z\downarrow>_e|\text{Box A untriggered, Box B triggered, Erwin alive}>.$$

Putting Erwin in Box B will yield

$$\sqrt{\frac{1}{3}}|z\uparrow>_e|\text{Box A red triggered, Box B untriggered, Erwin alive}>+$$

$$\sqrt{\frac{1}{3}}|z\uparrow>_e|\text{Box A green triggered, Box B untriggered, Erwin alive}>+$$

$$\sqrt{\frac{1}{3}}|z\downarrow>_e|\text{Box A untriggered, Box B triggered, Erwin dead}>.$$

The argument now proceeds by invoking symmetry principles and indifference principles. If all you care about is the health of Erwin's successor, then whether a successor happens to be in Box A or Box B alone makes no difference to the value of an outcome, and similarly whether the red or green trigger, or neither, happens to be triggered. But if we ignore the box and trigger labeling, all six decoherent states in the two outcomes are identical in all relevant respects, even in amplitude, except for whether the Erwin-successor is alive or dead. If you put Erwin in Box A, his successors end up dead in two of the three equivalent branches; if you put him in Box B, only one of his successors ends up dead. So by a dominance principle, you should prefer putting him in Box B. The end result will be more live Erwin-successors, where all the successors must be regarded as equally valuable.

By appealing to a symmetry principle and a dominance principle, then, one can argue that putting Erwin in Box B is the rational action to take. Wallace's proof parlays this sort of strategy into a general form, covering all sorts of decision situations. This is not a full presentation of the argument, but it gives the flavor of how it works.

So far we have not touched our original problem at all: how to make the choice in the original set-up. What we now need are additional indifference principles entailing that the new decision problem must be regarded by a rational agent as *equivalent for all decision-theoretic purposes* to the original. If we can show this, then the solution to the second set-up dictates the solution to the original.

Such indifference principles must render the "rational" agent indifferent to a lot of physical facts. For example, the branching structure in the second problem is unlike that in the original: Instead of branching into two decohering states, the new experiment branches into three. Further, the decohering macrostates in the two cases are themselves different. For example, the physical set-ups of the experiments at the beginning are not the same. Wallace invokes two "rationality principles"—Branching Indifference and Macrostate Indifference—which imply that no rational agent should consider these differences in the experiments relevant to the decision choice.[5]

There are other things that a "rational" Everettian agent is forbidden from doing. No rational agent can directly desire to create certain sorts of branching structures and value the existence of such structures more highly than the existence of either branch individually. Suppose, for example, you are presented with the choice of two delicious desserts at a three-star Michelin restaurant. You would like to try both, but both the cost and your present satiation make that undesirable. You value each equally, so there is no rational objection to picking one, picking the other, or flipping a coin. These are all options for the GRW agent or the Bohmian agent. But the Everettian agent has yet another option: Send a z-up electron through an x-oriented Stern-Gerlach magnet, and let the choice of dessert be conditional on the outcome. As an Everettian, you will foresee with certainly the outcome: The world will evolve into two decohering branches, on one of which your successor eats one dessert and on the other of which your successor eats the other. In a sense, you will get to have your cake and panna cotta, too. Would it be rational to pay for the experiment to be done, knowing that at least one of your successors gets to try each dessert?

Such a preference must be forbidden by the rationality principles, since it inspires an action that would not be considered rational in traditional decision theory. In the traditional setting, there is no sense in which it is possible to get a "both desserts" result: No matter what you do, you will only eat one. Paying for the experiment is irrational: You pay the cost, so any free choice would have given an overall better outcome. So an Everettian who prefers to pay for the split "both/and" outcome in preference to any "either/or" outcome will be irrational by the usual decision-theoretic standards. Since Wallace wants the rational Everettian to behave normally, this preference structure must be forbidden. In this case, the technical work is buried rather deeply

[5] Wallace (2012), p. 179.

in the axioms, in the definition of what can possibly count as a "reward."[6] The set of possible rewards (i.e., the things assigned intrinsic values) must all be mutually orthogonal in Hilbert space, so if eating cake is a possible reward and eating panna cotta is possible reward, then the branching structure in which different things are eaten on different branches cannot be a reward, cannot be aimed at or valued as such. Nothing is said about *why* one can't rationally value such an outcome. It is just built into the technical machinery that one can't.[7]

We can now see one main conceptual issue with this decision-theoretic approach. Granting that the theorem is correctly proven, what has been shown is that an agent who satisfies some constraints that are given the title "rationality axioms" and who makes choices in a situation restricted by a set of conditions about what can count as a reward must make decisions that are identical to those an agent following standard decision theory in a setting of uncertainty. The latter agent is unsure about which outcome will occur in the future and assigns the probabilities via Born's Rule. Wallace contends that getting the rational Everettian agent to behave the same way counts as solving half the challenge confronting Many Worlds, the half he calls "the practical problem." The practical problem is to show that a rational Everettian must act as if the squared amplitudes of future branches were probabilities, in the sense of making the same decisions as a standard agent (i.e., an agent maximizing expected utility in a state of uncertainty) would. But there are at least two further issues. One is whether the structural constraints put on the decision-theoretic set-up (e.g., not regarding a certain branching outcome as itself intrinsically valuable, being indifferent to branching structure)

[6] Wallace (2012), p. 175.

[7] Wallace discusses this objection but offers a bad analogy to suggest that standard decision theory faces similar issues. He likens it to an agent who pays to have a decision be settled by the flip of a coin "because, say, he finds it comforting to have the decision taken from his hands; the reader can probably supply other motivations" (Wallace 2012, p. 194). But this is not a case of preferring that a decision be made a certain way but of preferring that there be a different sort of outcome than can occur in the classical theory.

are really rationally compelling. The other is whether, even granting the theorem, one has gone any distance toward understanding how Many Worlds theory can recover the usual understanding of the implications of Born's Rule.

Here is an indication that the latter issue has not been resolved. Consider again your choice for which box to put Erwin in. Standard decision theory implies that in a case of uncertainty about the outcome but accepting the Born's Rule probabilities for the possible outcomes, a rational agent ought to put Erwin in Box B but also be unsure if that choice will yield the desired outcome. Wallace's theorem shows that a rational Everettian must make the same choice. But in a GRW world or in a Bohmian world, the choice of act is accompanied by foreboding: If the less likely event should occur, and Box B gets triggered, you will have failed to save Erwin. The Everettian has nothing obvious to be worried about. The actual branching outcome, with the various squared-amplitude weights, is known for certain in advance. So the "rational Everettian" acts like an agent faced with uncertainty in that she makes the same choice, but unaccompanied by foreboding.

If the game, then, is to show that making choices in an Everettian branching multiverse should be just like choosing in a GRW world or a Bohmian world, that task has not been accomplished. The Everettian only acts like an agent assigning Born Rule probabilities to exclusive alternative outcomes in one respect: the actual choice made. But the significance or meaning of the choice is different, and the difference should show up in one's emotional state and attitudes. Again, it is not clear why the Everettian should even be interested in showing that accepting an Everettian multiverse should make no difference to how one regards one's choices and their outcomes. Offhand, one might think that accepting the existence of Many Worlds and branching should have all sorts of ramifications for how one regards oneself, one's choices (including the range of choices), and the world as a whole. But modern Everettians have tended to eschew this idea and have gone to great lengths to minimize the practical effects of accepting this remarkable view.

It is even contentious to assert that the Everettian agent "acts like" a standard agent in a situation of uncertainty. It is at least as correct to say that the rational Everettian agent acts like a standard agent in a condition of certainty about the outcome of various choices, but who also values the future decohering branches in proportion to their squared amplitudes. Such an agent would prefer to put Erwin in Box B. She is sad that the Erwin-successor on the branch in which Box B is triggered will die, but she cares less about that Erwin-successor just because the squared amplitude of the branch is lower than the squared amplitude of the branch on which the Erwin-successor lives. Standard decision theory operates by calculating an expected utility for each act, weighting the value of each outcome by its probability and summing over all possible outcomes. The same number is produced if one simply revalues an outcome proportional to the squared amplitude of each branch and sums these values on the grounds that all branches will be created. The decision problem is converted to decision-making under certainty rather than uncertainty, and given the revaluation of the outcomes, the agent makes the same choices as the uncertain agent using probabilities. But subjective uncertainty, and its accompanying anxiety about the outcome, disappear.

Hilary Greaves (2004) argues that even if an Everettian agent makes choices that match those of a subjectively uncertain agent, it is not correct to say that the Everettian is subjectively uncertain about anything. After all, what is there to be uncertain about? She knows (well enough) the present wavefunction of the lab, and she knows the branching structure that decoherence of the wavefunction will create. There is no physical fact not determined by the wavefunction, so the physical state in the future is completely foreseeable. At the time of the choice, the agent is not uncertain about which future Erwin successor is "really" Erwin: they both will have equal claim to that title. Nor is she uncertain about which of her successors will be "the real" her, for the same reason. Greaves argues that these decision-theoretic arguments can show only that the rational Everettian must adopt a certain *caring measure* over future outcomes: The agent must care more about events in high-amplitude branches and care in proportion to the squared amplitude.

In sum, modern Everettians have theorems that a rational Everettian agent must make choices just like a rational agent faced with uncertain outcomes, where the likelihood of the outcomes is given by Born's Rule. But there are several problems still to be faced. One is that both the way the decision-theoretic situation has been constrained and some of the rationality principles seem unmotivated. The other is that even granting the theorem, it has not been shown that accepting the Everettian picture should not radically alter one's understanding of what choosing is and what the consequences of one's choices might be. In that sense, one has not recovered the standard Quantum Recipe.

Solving the practical problem of rational decision-making would not, in any case, vindicate all everyday uses of probabilistic concepts. Probability also plays a central role in how we evaluate some claims as evidence for other claims. Here is a case similar to one discussed by Wallace.[8] Suppose that a beam of electrons is coming into a lab, and one knows that either the electrons are all spin-up in a direction 5° away from the z-direction or in a direction 10° away from the z-direction. Suppose further that the only instruments one has to hand are a z-oriented Stern-Gerlach magnet and a mechanical device that registers whether an electron is deflected up or down. One wants to accumulate evidence that will help decide between the two hypotheses about the direction of spin of the incoming beam.

The solution is easy. Set the beam to run through the Stern-Gerlach apparatus and turn the counter on. According to the usual application of Born's Rule, the chance that an individual electron in the 5° spin state will be deflected up is $\cos^2(2.5°) = 0.998$. The chance that an individual electron in the 10° spin state will be deflected up is $\cos^2(5°) = .992$. If the experiment is left running all night and 1,000,000 results are tallied, then the two hypotheses make the following probabilistic predictions. If the beam is spin-up in the 5° direction, there is a 99.99% chance that

[8] Wallace (2012), pp. 201ff.

between 997,830 and 998,170 of the electrons will be recorded as deflected up; whereas if the beam is spin-up in the 10° direction, there is the same probability that between 991,660 and 992,340 of the electrons will be recorded as deflected up.

The next day when the data have been collected, it seems obvious how to proceed. If the data show a frequency of upward deflections between 997,830 and 998,170, accept the 5° hypothesis and reject the 10°. If the data show a frequency between 991,660 and 992,340, reject the 5° and accept the 10°. There is a chance that this conclusion will be wrong but 1) the chance is small, and 2) acceptance and rejection of hypotheses in this way is known to be extremely reliable in cases where the correct hypothesis can be independently verified. The basic epistemological move here is an application of conditionalization using Bayes' Theorem: If two alternative hypotheses ascribe different probabilities to a phenomenon, then the hypothesis ascribing the higher probability is better confirmed by the occurrence of the phenomenon. There are many subtleties about the exact use and justification of this sort of inference, but as a historical fact, it has been remarkably reliable.

In cases like this, we behave and reason as if the possibilities that are ascribed probabilities very close to 0 do not occur and possibilities ascribed probabilities close to 1 do. (This characterization is rough and requires refinement, but it serves as a good first approximation.) If the relevant high-probability outcome occurs, the procedure outlined above selects correctly between the two hypotheses. And the same condition also vindicates classical decision-making recommendations if the decision is made repeatedly.

Why should one follow the standard decision-theoretic advice to maximize expected utility? After all, what we desire to maximize is not expected utility but utility. This is the source the standard agent's foreboding discussed earlier in the chapter: Having made the decision about which box to put Erwin in, the agent knows that expected utility has been maximized, but not what the actual utility of the outcome will be. But if we confront the same decision situation over and over, making the same choice each time, it becomes more and more probable that the average utility we receive per decision will be very close to the calculated expected utility. So if this high-probability outcome happens, the effect of following the standard advice is to maximize our long-term profit relative to always having made some other decision. This observation does not hold for individual decisions or shorter sequences of decisions, because the probabilities for these decisions never approach 1. Individual gamblers sometimes make great profits and sometimes suffer great losses, but casinos regularly and predictably make money. The reliable effectiveness of standard decision theory for casino managers can be accounted for by the very same fact that accounts for the effectiveness of the evidential inference discussed above: If the relevant very-high probability outcomes occur, casinos will turn a profit and scientists will reach the correct conclusion.

For a collapse theorist or a pilot wave theorist, then, the practical problem of decision theory and the epistemic problem of evaluating evidence are linked. The same physical condition implies the success of both. But the same cannot be said for the Many Worlds theorist. Since all decohering branches actually will exist, the long-term profitability of a decision rule cannot be accounted for by the nonexistence of low-amplitude branches. And the evidential problem also takes a different form. In Many Worlds, it is not possible for the inference made above to always or usually be reliable. It will be reliable on some branches and unreliable on other branches in every case. The branches on which it is reliable will have a relatively higher squared amplitude than the branches on which it fails. The Many World theorist therefore seeks a normative argument to the effect that we *ought* to form our beliefs as would be appropriate for the higher-amplitude branch.

In Many Worlds, the practical problem, which is oriented to the future, is addressed by a rule concerning how strongly we should value various future outcomes. Even if this advice is accepted, such a normative suggestion has no obvious bearing on the epistemic problem. Instead, the epistemic problem requires us to make inferences that are appropriate only if we are already on branches with relatively high squared amplitudes.

When Wallace discusses this problem, he finesses the difference between the future-oriented problem and the past-oriented problem. This is done by asking not what the agent getting up the next morning in our spin-experiment example should infer from the collected data, but what the prebranching agent the night before should commit to doing. The prebranching agent foresees having future successors on high-amplitude branches who make good inferences, and future successors on low-amplitude branches who make bad inferences. So Wallace appeals to the practical argument: The prebranching agent neglects the future branches with bad inferences, because they are low-amplitude. So she doesn't care about them.[9] But the postbranching agent cannot argue this way. That agent only cares about getting the right conclusion for herself. She can't dismiss the importance of making an error by saying, "well, if I am the low-amplitude successor, then my success doesn't really matter."

The next morning, having recovered the data, the Many Worlds theorist has to decide what to infer. The collapse theorist or the pilot wave theorist in the same situation can reason, given the physical hypothesis that the high-probability outcome occurred, that the data from the lab reflects which hypothesis was correct. But the Many Worlds theorist knows that no matter what results come in from the lab, those results were certain to occur on either hypothesis. What is sought is a normative rule that dictates believing that her own branch has a relatively high squared amplitude rather than low one, even though the squared amplitude makes no qualitative difference to the branch.

What is conceptually jarring about Everettian approaches to this problem is this normative flavor. These approaches attempt to show that various rationality principles require that an agent *ought to* adjust her beliefs in certain ways. But no matter how these normative matters come out, there is a non-normative fact to explain. These techniques for forming beliefs and making choices actually have worked well. This is not a normative fact in itself. This fact plays a role in explaining why animals using

[9] Wallace (2012), p. 202.

certain strategies have evolved, while others died out. It is hard to see how *normative* constraints bear on this. The animals were not rational, and their behavior was not dictated by their appreciation of any norms. Even if a decision strategy violates the norms, that alone cannot explain why the animal failed to evolve unless one explains why violating the norms leads to bad consequences. But the normative character of the rules is irrelevant to that question.

Collapse theories and pilot wave theories can cite a physical characteristic of the universe that, if it obtains, accounts for the utility of the practical and epistemic advice. The same characteristic would also explain why evolution has proceeded as it has. If certain sufficiently low-probability events do not occur, the advice yields good results (in the long term), and only certain behaviors will be successful. But no parallel physical fact about a Many Worlds universe can guarantee the success of the advice. It will always work for some and fail for others. The normative principles endorse strategies that work for the relatively high-amplitude sucessors and fail for the relatively low-amplitude ones. We can justify using the rules to make practical decisions if we decide to care more about our high-amplitude successors than our low-amplitude ones, but how could our decision about what to care about in the future have any bearing on the course of evolution in the past, or the historical success of inferences such as those about the electrons? It is hard to see how to integrate the normative character of these proofs with the physical facts we seek to explain.

The Problem of Local Beables and Macroscopic Reality

In one sense, our discussion of the Many Worlds theory has mirrored our discussion of GRW: A dynamics for the quantum state is specified via a dynamical equation for the wavefunction, and various consequences of that dynamics are noted. In the case of GRW, only after that analysis did we observe that if the total ontology of the theory is just the quantum state, then there are no local beables, and it is obscure how to connect the behavior of the quantum state alone to the sort of data (about the macroscopic situation in laboratories) that we take to report actual physical events. We further have seen that the GRW quantum state dynamics can be allied with different sorts of local beables, yielding different physical theories. So the time has come to ask the parallel question for Many Worlds. What, if anything, are the local beables in this theory, and how does the basic physical ontology of the theory connect to the sorts of facts that are accepted as data?

Just as with the GRW theory, this problem has been obscured by a linguistic labeling trick. For example, we have asserted that, by virtue of the linear dynamics, the state $|x{\uparrow}>_e|\text{ready}>_a$ will evolve into $\frac{1}{\sqrt{2}}|z{\uparrow}>_e|\text{points up}>_a + \frac{1}{\sqrt{2}}|z{\downarrow}>_e|\text{points down}>_a$. And the way we have spoken about the quantum state suggests that $|\text{ready}>_a$ (which is a wavefunction and nothing else) represents a physical situation in which some macroscopic apparatus is in a "ready" state of some sort, $|\text{points up}>_a$ represents a physical situation in which that macroscopic apparatus has gone into one macroscopic indicator state, and $|\text{points down}>_a$ represents a physical state in which it has gone into a different indicator state. After all, if I call a quantum state "$|\text{points down}>_a$," it must surely be a state in which (in the usual sense) a macroscopic apparatus indicates a "down" outcome. But by what rights is the label justified by the structure of the physical state it labels?

When analyzing both collapse theories and the pilot wave theory, we solved this problem via local beables. If there are particles, or flashes, or a continuous matter density distributed through space-time, then we know how to proceed. The precise microscopic matter distribution in a model determines the macroscopic situation by simple aggregation. Applying the theory to the entire laboratory situation yields empirical predictions. In contrast, if the theory postulates no local beables, it is not clear how to proceed.

One Many Worlds approach to this question is to deny that familiar macroscopic objects need to be composed of local microscopic parts. Instead the macroscopic objects of familiar experience and of lab reports are to be analyzed functionally as patterns in the behavior of the quantum state. If the quantum state evolves in a certain way, it is said, a cat thereby comes into existence with no additional physical ontology needed. Indeed, the whole low-dimensional space-time itself somehow emerges from the behavior of the quantum state.

We have met this sort of approach before, in our discussion of emergence in the GRW theory. We have already seen the difficulties involved in defending the view that "to be a table or a chair or a building or a person is—at the end of the day—*to occupy a certain position in the causal map of the world*." [10]

When expositing Many Worlds, David Wallace adopts a similar approach. Tables and chairs and people are, according to Wallace, *patterns*. "[A] macro-object is a pattern, and the existence of a pattern as a real thing depends on the usefulness—in particular the explanatory power and predictive reliability—of theories which admit that pattern in their ontology."[11] The implicit claim is that the sorts of patterns that constitute tigers can be instantiated in the behavior of quantum states, even if the quantum states themselves have no obvious relation to any familiar space-time.

Wallace does not lay out this situation bluntly. Shortly before the passage just cited, he says this: "The moral of the story is: there are structural facts about many microphysical systems which, although perfectly real and objective (try telling a deer that a nearby tiger is not objectively real), simply cannot be seen if we persist in describing those systems in purely microphysical language."[12] But in this passage, the point is uncontroversial. Tigers are typically understood as complicated systems with a definite microscopic structure. What makes a collection of electrons, protons, and neutrons a tiger, rather than something else, has to do with how the microscopic parts are arranged, or structured. A significant part of that structure is the spatial arrangement of the microscopic parts through time. That much already determines the size, mobility, dental sharpness, and so forth, of the tiger. The

[10] Albert (2014), p. 217.
[11] Wallace (2012), p. 50.
[12] Wallace (2012), p. 50.

task of seeing how these macroscopic features of a tiger follow from the structuring of its microscopic parts is conceptually clear. But a quantum state contains no microscopic parts localized in space-time, so its behavior—whatever it is—cannot create a macroscopic tiger in anything like the way the behavior of localized electrons, neutrons, and protons can. In what sense, then, is a tiger the kind of pattern that could even in principle be instantiated by a quantum state?

There is no standard answer in the Many Worlds literature. Indeed, the Many Worlds approach often takes for granted that "patterns" or "behaviors" or "causal structures" in the quantum state alone can unproblematically be familiar macroscopic objects, even though the quantum state is not a local object in a familiar space-time. Whatever patterns these are supposed to be, they are not the patterns created by the behavior of microscopic physical entities in space-time.

This assumption is critical for the basic argument for Many Worlds. We assume that some structure of a quantum state can constitute the physical situation we know familiarly as "apparatus in the ready state," another structure can constitute the situation "apparatus indicating up," and yet another the situation "apparatus indicating down." Assuming this, it is appropriate to label one quantum state $|\text{ready}>_a$, another $|\text{points up}>_a$, and yet another $|\text{points down}>_a$. But then it is hard to resist the Many Worlds conclusion: If the universal quantum state evolves into $\frac{1}{\sqrt{2}}|z\uparrow>_e|\text{points up}>_a + \frac{1}{\sqrt{2}}|z\downarrow>_e|\text{points down}>_a$ and the two branches decohere, the behavior of the squared amplitude of the universal quantum state is a sum of the behaviors of the squared amplitudes of $|z\uparrow>_e|\text{points up}>_a$ and $|z\downarrow>_e|\text{points down}>_a$. Because of the decoherence, the patterned behavior of the squared amplitude of $\frac{1}{\sqrt{2}}|z\uparrow>_e|\text{points up}>_a + \frac{1}{\sqrt{2}}|z\downarrow>_e|\text{points down}>_a$ contains within it (at smaller squared-amplitude scale) both the pattern produced by $|z\uparrow>_e|\text{points up}>_a$ and the pattern produced by $|z\downarrow>_e|\text{points down}>_a$. So if all it takes for there to be a z-spin up electron and a "points up" apparatus is for one pattern in the squared amplitude to exist, and all it takes for there to be a z-spin down electron and a "points down" apparatus is for the other pattern to exist, then there are now two apparatuses and two electrons, in different states.

This argument is so powerfully lodged in the Everettian mind that it can blind one to the structure of alternative theories. David Deutsch has opined that "pilot-wave theories are parallel-universe theories in a state of chronic denial."[13] The notion is that the linearly evolving quantum state alone already implies the existence of a multiplicity of familiar macroscopic worlds. Since Bohmian mechanics posits such a quantum state, the addition of actual particles on particular trajectories could at best add yet another, superfluous, world. This misunderstands the role of the particles—the local beables—in that theory. If tables and chairs and cats are structured collections of local entities in space-time, then stripping the local entities from the theory strips out all the familiar material objects as well. Furthermore, material objects made of Bohmian particles, or GRW flashes, don't have the same conceptually problematic features as decoherent branches of the quantum state: They don't split or divide, objects do not have multiple successors, and so on. These theories face fewer conceptual problems explicating probabilistic language: The probabilities are for the local beables to be arranged one way or another.

Insofar as there are no obvious local beables in the Everettian ontology, it is unclear how to connect the ontology of the theory to the everyday world. Wallace and Chris Timpson (2010) have made a proposal addressing this problem, called "Spacetime State Realism." One aim of the theory is to explain how the Everettian can, after all, attribute physical contents to regions in space-time.

Wallace and Timpson frame their account using field theory. Field-theoretic states, including the quantum state, are specified using mathematical objects tied to regions of space-time. If we take this at face value as implying a commitment to space-time itself (how else to understand the regions?), then the ontology of the theory already contains both the universal quantum state and a familiar space-time structure. But positing a space-time is of little use if there is no matter localized in it.

[13] Deutsch (1996), p. 225.

Given the universal quantum state, there is a natural way to define a mathematical object associated with a limited space-time region, called the *reduced density matrix*. The basic idea behind space-time state realism is to treat this reduced density matrix as representing a real part of the physical ontology, a part localized in the given region. The mere existence of the mathematical object alone does not magically imply the existence of any such ontology, just as the collapses of the GRW wavefunction alone do not imply the existence of the flash ontology. The mathematical function from collapses to space-time points exists willy-nilly, but the theory only comes to have physical flashes at those points via an act of physical postulation. Similarly, the mere mathematical existence of a function from the wavefunction to a mathematical density in space-time does not imply the real existence of a matter density. And the mathematical existence of "Bohmian trajectories" definable from the wavefunction via the guidance equation does not imply the real existence of Bohmian particles. The flashes or the matter density or the particles only get into the physical ontology as physical postulates of the theory. Once there, their behavior can be described by these mathematical functions. But the functions alone, which exist of necessity as mathematical objects, do not call them into existence. Otherwise, all these ontologies would automatically follow from the existence of the quantum state!

So one could fill out an Everettian physical ontology as follows. In addition to a quantum state, there is a real, macroscopically familiar space-time structure, and in that space-time, there is a physical magnitude described by the reduced density matrix. We have inflated the ontology beyond just the quantum state but now have a familiar space-time and something in that space-time to show for it.

Several challenges ensue. One is this: Whereas the distribution of particles in space-time in a Bohmian theory can correspond in an obvious way to the world we think we see, and the distribution of flashes in GRW with a flash ontology, and the distribution of high-density matter in GRW with a matter density ontology can as well, the Everettian reduced density matrix presents nothing like a picture of the world we think we live in. Instead, intuitively, the reduced density matrix at best corresponds to all the distributions of matter on all the decoherent branches superimposed on one another. Given how much branching and chaotic mixing of matter there has been since the Big Bang, the reduced density matrix of any region of space-time would be almost completely uniform at all scales. It will display no notable structure at all.[14]

It is here that appeal is made to branching. The reduced density matrix of individual decohering branches might show a familiar spatial structure—perhaps one could pick out regions where stars are, and planets, and trees, and so on. But the branching structure in Many Worlds is only approximate rather than rigorously defined. So the phrase "the density matrix of a region relative to a branch" does not have sharp mathematical content. Only "the density matrix of a region relative to the complete universal wavefunction" (or "quantum state") does.

Ignoring the fact that branches are at best only vaguely defined, would the reduced density matrix of a region, relative to a branch, correspond in any obvious way to what we take the physical situation in the region to be? We cannot say without a specific theory—and hence a specific density matrix—to analyze. But Spacetime State Realism is more a programmatic suggestion rather than a precise theory, so it is hard to tell. For example, one case Wallace and Timpson discuss involves spin interactions between particles that are *stipulated* to have definite space-time trajectories.[15] The trajectories are required to determine the interaction Hamiltonian between the particles. So these examples provide no clue as to how to get the trajectories in the first place. But ultimately, the trajectories are needed to connect theory to data.

In the standard theory, these density matrices are not used to characterize the intrinsic structure of matter. They are used

[14] An anonymous referee objects that for small regions, the reduced density matrix will be structureless; for middling-sized regions, there can be lots of structure of entanglement. But after 13.7 billion years of interaction, the entangled structures will be at such a huge scale that restricting to any everyday scale will lose the entanglement due to entangled areas being left outside.

[15] Wallace and Timpson (2010), p. 723.

instead to calculate probabilities for measurement outcomes. So the standard use of these mathematical objects already presupposes a solution to the measurement problem, which is exactly the thing we are now trying to use them to solve. What happens if we try to take them instead as representing the intrinsic structure of some localized matter?

A curious feature of positing a real physical quantity corresponding to the reduced density matrix is that although each region has a density matrix associated with it, the quantity so defined is not *locally separable* in the way that particles or flashes or matter densities are. Local separability requires the following feature: Specifying the locally separable ontology in each of two regions suffices to specify the locally separable ontology in their union. In other words, the characteristics of a large region supervene on the locally separable characteristics of its parts. The global distribution of particles (or flashes, or mass density) is determined by the microscopic local distribution. Specify what there is in a collection of small regions that cover the whole space, and you specify the distribution for the whole space. But density matrices don't work like that. The density matrix associated with the union of two regions typically contains more information—more structure—than the pair of density matrices of the regions. The source of this is entanglement between the regions.

Failure of local separability reverses the usual relation between part and whole. Since in a locally separable ontology, fixing the states of the parts fixes the state of the whole, it is natural to regard the parts as metaphysically more fundamental than the whole: The whole is "just the sum of the parts." This inversion of the usual relation between spatial parts and wholes means that we cannot infer the macroscopic situation—the situation in a large region—from the state in the microscopic parts.

Spacetime State Realism, then, cannot follow the same strategy for connecting theory to data as do the other theories we have examined. Those theories postulate a sharply defined unique distribution of local beables at microscopic scale whose collective behavior implies how macroscopic objects are shaped and move. This, in turn, can be compared to the macroscopic behavior of laboratory apparatus. At the microscopic scale, all Spacetime State Realism can say about the local material content of regions is that it is nearly uniform everywhere: The decoherent branching structure will not be defined at that scale. The macroscopic situations we want to compare to laboratory reports cannot be recovered from the microscopic conditions. The Many Worlds theorist is forced to discuss the situation at macroscopic scale without reducing it to microscopics. And the relevant discussion at the macroscopic scale—the structures that are supposed to correspond to data reports—will not be sharply defined, since the decoherent branching structure itself is not.

Can this yield a satisfactorily clear account? It is hard to tell without understanding how the details are filled in. The suggestion of Wallace and Timpson is programmatic rather than specific. And the specification would require providing the exact details of the field theory, an investigation of how the density matrix and decoherence structure of the theory behave at various scales, and the interpretive principles invoked to connect that behavior with macroscopic behavior.

Our decision to study nonrelativistic quantum mechanics has now come back to haunt us. That decision was pragmatic: The mathematical structure of nonrelativistic quantum mechanics is easier to present and learn, and so it allows us to investigate some characteristic features of quantum behavior in an admittedly simplified setting. Questions about the dynamics of the quantum state, the justification of probabilistic language, locality, interference, decoherence, and examples of different sorts of local beables could be presented. The hope was that the nonrelativistic theory could at least provide a model for how properly relativistic extensions can be constructed. And one of my main themes is the methodological advantages of postulating local beables for making empirical sense of the theory. We have seen how particles, flashes, and matter densities might play this role in the nonrelativistic regime.

Having gotten to this far, we can go no further. Wallace and Timpson recognize the methodological utility of having a description of something distributed in space-time when trying to understand the theory. They suggest using the reduced density matrix associated with space-time regions in field theory as the mathematical representation of that "something." If we take the step of making a commitment in the theory to this additional ontology, we have something to work with to explicate the physical structure of familiar objects. But we have traded the exact-but-unrealistic simplifications of the nonrelativistic theory for the more-realistic-but-programmatic discussion that appeals to field theory.

At least we can say this: Spacetime State Realism is not an ontologically monistic theory. It postulates a quantum state of the universe, and in addition a space-time structure, and in addition a local-but-nonseparable physical content to regions of the space-time. Despite Wallace's appeal to a very generic form of functionalism, it is ultimately the branch-dependent behavior of this localized ontology that might allow us to make contact with the language of data. It would be in terms of the local state of regions of space-time that we would understand claims about how macroscopic objects are structured and move. But what that local content precisely is, and hence how it behaves, cannot be discussed at this generic level. We would need a detailed field theory to make further progress.

Further Reading

The literature on Many Worlds is vast. An excellent collection of discussions of the theory, both critical and supportive, is Saunders et al. (2010). Discussion of the theory has evolved. An overview from an earlier perspective is Barrett (1999).

CHAPTER 7

Relativistic Quantum Field Theory

THE LAST THREE CHAPTERS have discussed different sorts of exact theories designed to recover the predictions of the Quantum Recipe for nonrelativistic quantum mechanics, including the predictions for our original eight experiments in chapter 1. The exposition of these theories has taken a classical, nonrelativistic space-time structure for granted. An absolute notion of simultaneity is presupposed in the definition of the configuration space of a system, and hence in the definition of the wavefunction as a complex (or spinorial) function on configuration space. The exact dynamics of the original GRW theory employs the notion of simultaneity in specifying the collapse dynamics, and the guidance equation in Bohmian mechanics makes essential use of the objective time order of distant events. As we have seen, in Bohmian mechanics, the exact time order of two distant spin measurements on an entangled pair of particles can determine whether an experiment yields one observable outcome rather than another. When measuring the z-spin of a pair of particles in the singlet state, for some particular initial conditions, one will get the result up-on-the-right-and-down-on-the-left if the right-hand experiment is performed first and the result down-on-the-right-and-up-on-the-left if the left-hand experiment is performed first. "First" in the preceding sentence refers to the order of the experiments in an objective absolute time structure.

The theory of relativity, however, denies the existence of any such objective time structure. In that theory, experiments carried out far apart (at spacelike separation from each other) have no objective time order: neither occurs "first" or "second." Constructing a completely relativistic theory along pilot wave lines is therefore a nontrivial task. Similarly, an objective collapse theory must be formulated differently in a relativistic context, since it no longer makes any sense to say that the collapse is instantaneous. The Many Worlds approach does not fall prey to either of these difficulties, since the quantum state never collapses and there is no additional guidance equation.

So one challenge facing us is adapting an exact quantum theory to a relativistic space-time domain. Yet another challenge arises from physical phenomena not illustrated in our eight experiments.

The extension of the mathematical techniques of the quantum formalism to a relativistic setting implies the existence of antiparticles, such as the positron. (A positron is physically identical to an electron except that it has the opposite electric charge.) Experimentation reveals that sufficiently energetic interactions can produce electron/positron pairs, so at the end of the experiment, there are more particles than at the beginning. In our eight experiments, nothing like this happens: I described those experiments in terms of the behavior of a fixed number of particles. Since the configuration space of the system depends on the number of particles, having a variable number of particles requires further adjustment to the mathematical formalism. Indeed, many physicists would insist that in this setting, it is misleading to speak of "particles" at all. Rather than talking about electrons, they say, one should speak of "the electron field." In certain experimental conditions, this field can produce behavior reminiscent of classical particles, but in other conditions, no such picture is available. One can define a "number operator" acting on this field that has eigenstates with different values. These are interpreted as quantum states with various numbers of particles. But if the state of the system in not an eigenstate of this operator, then there is no definite number of particles, and "particle talk" is not appropriate for the system.

At the least, then, the nonrelativistic Quantum Recipe must be modified to accommodate the account of space-time structure presented in the theory of relativity, and also to accommodate the experimental phenomenon of particle creation and annihilation. The formalism designed to meet these challenges is called "quantum field theory" (QFT), and it is in this mathematical language that our most exact and predictively accurate physical theories of matter have been formulated.

One might contend that the attention we have paid to nonrelativistic quantum mechanics so far has been misplaced. If the currently best predictive formalism is QFT, then our interpretive efforts ought to have been directed at it from the outset. For several reasons, I have not proceeded in this way.

First, QFT is much more mathematically complex than quantum mechanics. Since many iconic quantum phenomena—two-slit interference, quantization of spin, EPR correlations, violations of Bell's inequality—already appear in the simpler setting, it is a pedagogically useful place to begin. We were also able to more easily discuss the basic strategies for interpreting the formalism, namely, postulating an objective collapse of the quantum state, adding additional variables, and trying to do without either of these in a Many Worlds approach. We now face the prospect of modifying these theories to fit the relativistic context, but we can be guided by these examples of the basic moving parts in each sort of theory.

Second, the nonrelativistic Quantum Recipe does not suffer from certain mathematical difficulties that QFT faces. Calculations in that theory are carried out by summing over a sequence of terms, but if one allows the sum to cover all possible terms in the sequence, then the mathematics sometimes breaks down. In particular, calculations intended to produce probabilities for certain outcomes yield divergent sums—infinities—instead of finite answers. This problem led to the mathematical technique called "renormalization" and to a general outlook on QFT as an "effective theory." The basic idea is that the theory actually used to make calculations does not pretend to be complete or exact. Instead, it is a good approximation that holds so long as the energies involved in the experimental situation are sufficiently low. Instead of calculating predictions from an exact fundamental theory, one makes observations at a certain energy scale to fix effective constants for that scale, and then uses those constants to make further predictions in that energy regime, cutting off higher-energy terms from the sequence. Regarding the mathematics being used as that of an effective theory concedes that one is not giving a complete description of all the physics there is, but instead just the part of the physics most important at that scale.

Some physicists and philosophers are dissatisfied with this sort of theory as the basis for investigations into physical ontology, preferring a sharply defined mathematical formalism. One such approach is known as *Axiomatic Quantum Field Theory*. As a purely sociological fact, few physicists demand this sort of axiomatization. (But, of course, few physicists demand clear solutions to the measurement problem, or an account of the collapse of the wavefunction, or clarification of the other foundational problems we have been concerned with!) So individuals making foundational inquiries into QFT are immediately confronted with a choice: Should one focus on the sort of effective field theory that physicists use in day-to-day calculations, or instead begin with the axiomatic theory, even if it is not the sort of thing from which actual testable consequences have been derived? The philosophical community is not in agreement.

Assessing this dispute requires more technical detail than can be provided at our level of exposition, but readers interested in the issue can begin with papers by David Wallace (2006, 2011) and Doreen Fraser (2009).

Rather than attempting to address this issue, then, we will content ourselves with just the two already mentioned. First, what challenges does the move from a classical space-time with an absolute notion of simultaneity to a relativistic space-time pose for our different sorts of quantum theories? And second, what implications does the phenomenon of particle creation and annihilation have for proposed local beables of the theories?

Simultaneity and Foliations

Classical space-time structures postulate absolute simultaneity, that is, a temporal structure in which two events either happen at the same time or one precedes the other in time.[1] The collection of events that happen at the same time form a *simultaneity slice* of the space-time, and the whole space-time is *foliated* into a collection of such slices. That is, the space-time is carved, in a unique way, into a sequence of global leaves composed of simultaneous events.

In a classical setting, the configuration of a system is defined by reference to such a foliation. The configuration of a collection of particles at a given time is determined by the positions of the parts of the system at that time. The configuration of a field at a given time is determined by the values of the field at different spatial locations at that time. Since the foliation partitions the entire space-time into slices and the configuration specifies the distribution of matter on a slice, the whole history of the particles or fields can be described by the sequence of configurations.

In relativity, the structure of space-time does not in itself define any such unique foliation. Events at "spacelike separation" (i.e., events that lie outside each other's light cones) have no determinate temporal order: One can neither say that one happens first nor that they happen at the same time.[2]

The absence of a unique foliation may or may not pose a problem when adapting a theory to a relativistic setting. If the theory makes no important use of the notion of simultaneity in its fundamental dynamical laws, then the absence of a foliation may present few problems. Special Relativity demonstrates that absolute simultaneity plays no essential role in the formulation of Maxwell's laws of electrodynamics. Since electromagnetic effects are never propagated instantaneously in that theory, there is no need for absolute simultaneity. One important consequence is that classical electromagnetic theory cannot predict violations of Bell's inequality for experiments carried out sufficiently far apart that light signals from the measurement made on one side could

[1] This section presumes that the reader is familiar with the theory of relativity. If not, might I suggest the companion volume to this one: Maudlin (2012)?

[2] Details about these different space-time geometries can be found in the companion volume to this one (Maudlin 2012).

not reach the other. Since we need to recover such violations somehow, reconciling quantum theory with a relativistic space-time structure is bound to be tricky.

In the case of our nonrelativistic quantum theories, we will focus on two interrelated features: how the theory accounts for the observable behavior of macroscopic objects and how it generates violations of Bell's inequality. Presuming that the behavior of the macroscopic objects is determined by the behavior of their microscopic local parts, we investigate whether the time order of distant events plays a central role in the dynamics of those parts. If it does—as in the example above from Bohmian mechanics—then some fundamental adjustment must be made to adapt the theory to a relativistic space-time. One can distinguish three different strategies.

The first strategy requires finding relativistically well-defined replacements for the simultaneity structure and modifying the equations to use those in place of the classical foliation. While it is true that a relativistic space-time typically has no preferred foliation or "slicing," it does contain other sorts of geometrical structures not present in classical space-time. For example, the light-cone structure associates with each event a future and past light-cone. The interiors of the light-cones of an event can be foliated in a unique way by surfaces of fixed *proper time* from the event. These foliations are not global—they do not extend outside the light-cone—but for certain purposes, they may be sufficient for the task at hand. Whether they are sufficient depends on the theory we are trying to adapt.

The second strategy is to define a unique global foliation of the space-time somehow and then use that foliation in the relativistic theory in place of the foliation of absolute time in the classical theory. This global foliation could, for example, be determined by the distribution of matter as reflected in the quantum state. If the means by which the foliation is defined are themselves relativistically kosher (i.e., they do not make reference to any non-relativistic space-time structure), then the resulting theory can be defended as relativistic but can still make use of a global foliation in much the same way as the nonrelativistic version does.

The third strategy is simply to postulate a foliation as an intrinsic feature of the space-time. This does not mean returning to a classical space-time structure—classical space-times do not contain light-cones, for example. Supplementing a relativistic space-time with a unique foliation yields a new account of the spatiotemporal structure of the world. This is not a popular strategy—it is rightly seen as a rejection of relativity—but it nonetheless is viable. If a global foliation of the space-time is required to define the dynamical equations of the theory, this is an option to be explored.

How do our nonrelativistic theories fare with respect to these strategies?

The nonrelativistic GRW collapse dynamics makes use of simultaneity in specifying the collapses of the quantum state. In the flash ontology version, these collapses in turn determine the distribution of the local beables—the flashes—and hence the empirical predictions for the behavior of observable macroscopic matter. Regarded in this way, the collapses themselves play a somewhat instrumental role: What is ultimately important for the predictions of the theory is just a probability distribution over the possible distributions of flashes. If you know how likely it is that the flashes are distributed one way rather than another, then you know what the theory predicts empirically. The question then becomes whether the probability distribution can be generated without reference to any unique foliation of space-time, using only relativistic geometrical resources.

This is possible. Roderich Tumulka developed a version of the GRW collapse theory that makes use only of relativistic space-time structure for defining the probabilities for various distributions of flashes, using as initial conditions a quantum state defined on an arbitrary slice through space-time (a Cauchy surface) and a "seed flash" for each particle.[3] The seed flash determines both a future light-cone and a foliation of the future light-cone determined by the proper time from the seed. Since any subsequent flash for that particle must appear in the future light-cone, this foliation is sufficient to write down analogs to the nonrelativistic

[3] See Tumulka (2006).

collapse equations, using the foliation in place of classical absolute time.

Given these initial conditions, the theory makes probabilistic predictions. For example, given a pair of electrons in the singlet state in an experiment with two z-oriented Stern-Gerlach magnets, the theory ascribes a 50% chance for an up-flash-on-the-right-and-down-flash-on-the-left and a 50% chance for a down-flash-on-the-right-and-up-flash-on-the-left. The totality of predictions for all possible experimental arrangements predicts violations of Bell's inequality.

Tumulka's theory is nonlocal. Information about experimental outcomes far away (at spacelike separation) can improve one's predictions for local outcomes *even though everything in the past light-cone of the local experiment has been taken into account*. This is a consequence of entanglement of the quantum states of the two sides. (Contrast this with Bertlmann's socks, mentioned in chapter 1. Taking into account the past light-cone of the event of looking at one of his socks, one can predict with certainty what color will be seen, because the past light-cone contains the sock. Having taken this into account, information about the color of the other sock does not improve one's predictions.) The quantum states in Tumulka's theory show entanglement, and the theory violates locality, but it still requires nothing but relativistic spacetime structure for its formulation.

This theory, informally known as "Flashy Relativistic GRW," is not a complete physical theory. In particular, it does not provide for an interaction term in the Hamiltonian of the system. But the theory does provide a proof-of-concept example that nonlocality, evidenced by violation of Bell's inequality for events at space-like separation, can be implemented in a completely relativistic way.

Further investigations have produced fully relativistic matter-density theories with a GRW-like collapse dynamics. Once again, the dynamics is fundamentally probabilistic: From a given initial quantum state, different futures can arise. The collapse dynamics can be made relativistic in a way similar to that of Tumulka's theory. The specification of the matter density of a particle at a point in spacetime is postulated to be a function of the quantum state defined with respect to the past light-cone of that point. Since the light-cone structure is relativistic, this can be done without adding a foliation. This matter density does, however, behave in rather surprising ways. For example, starting with a "particle" in an equal superposition of traveling to the east and to the west, the matter density will contain equal-density lumps moving in both directions. If a single screen is set up far away in one direction, then the lump traveling that direction will continue until it meets the screen. The quantum state then evolves in one of two ways, with equal probability. One way corresponds to finding the particle at the screen: The matter density grows from half of the particle mass to its full value, and a mark is formed on the screen. The other way corresponds to not finding the particle at the screen, and the matter density is reduced to zero there.

The key question is what happens to the other lump, the one that has no screen to interact with. Since the matter density at a point is a function of the quantum state along its past light-cone, the other lump continues until the measurement event at the distant screen is in its past light-cone. At that point, the matter density on that side either increases to double its initial value (if the particle is not found at the screen; Figure 27b) or is reduced to zero (if the particle is found; Figure 27a). One is tempted to say that the quantum state "collapses along the future light-cone of the measurement event at the screen," although the disposition of the matter density in space-time is not so much due to how the quantum state behaves as to how the matter density at a point is calculated from the quantum state. By virtue of the use of the light-cone structure, the theory can be made completely relativistic. See Bedingham et al. (2014) for an overview of this sort of theory.

In sum, a GRW-like collapse theory can be formulated using only the spatiotemporal structure postulated by relativity, and this can be done for both the flash and the matter density ontologies. In both cases, the indeterminism of the theory plays a central role in achieving harmony with relativity.

Indeterminism allows these theories to respect all symmetries of the relativistic space-time. Consider again the case of a pair of separated electrons in the singlet state, subjected to distant

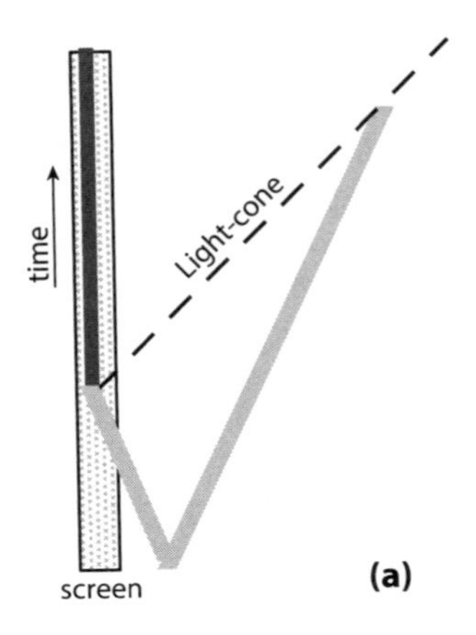

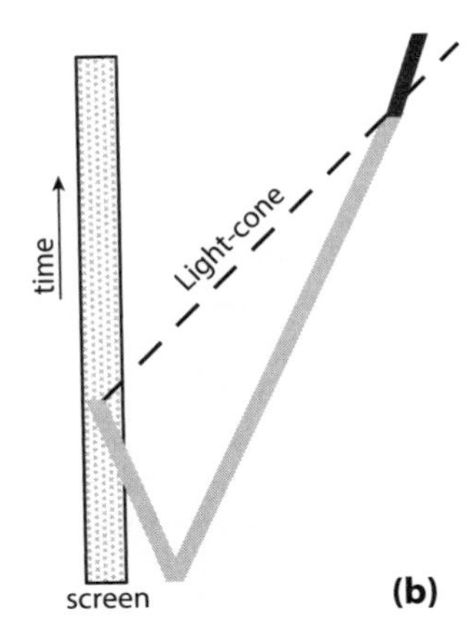

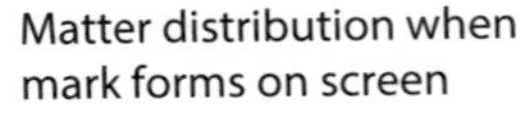

Matter distribution when
no mark forms on screen

Figure 27

z-spin measurements. The initial quantum state of the particles is completely symmetric between the two sides, so if the quantum state provides a complete physical description, then the electrons have the same physical characteristics. Nonetheless, the outcome of the experiment breaks this symmetry: On one side there is an "up" outcome and on the other "down." How does this symmetry get broken?

In an indeterministic theory, such as GRW, the symmetry is broken by sheer chance, by the collapse occurring one way rather than the other. The probabilities for the collapses can retain the symmetry, since each result can be ascribed the same 50% chance. So the probabilistic laws can respect the symmetry. In contrast, a deterministic theory must identify some physical feature of the experimental situation to break the symmetry and determine which outcome occurs. It is exactly here that tension with relativity occurs for the pilot wave approach.

As mentioned above, the classical foliation of space-time produced by absolute simultaneity figures in the guidance equation, so that the temporal order of distant experiments can influence the observable outcome. Attempts to formulate a relativistic pilot wave theory have therefore focused on the second approach: Using properties of the quantum state itself to define a foliation and then using that foliation in the guidance equation. This is discussed in, for example, Dürr et al. (2014). There are technical questions about the conditions under which a quantum state does determine a unique foliation (e.g., because of symmetries, a vacuum state in Minkowski spacetime cannot), but it is plausible that in any realistic situation, features of the quantum state can pick out a foliation. This raises the question of whether such a theory is "fundamentally relativistic." Since the foliation is determined from the quantum state, it is not implicit in the space-time structure independently of the matter. Nonetheless, any essential use of a global foliation when specifying the dynamics of a theory—even if the foliation itself is derived rather than fundamental—can easily strike one as contrary to basic relativistic postulates.

Part of this uneasiness about appeal to a global foliation is that the rejection of absolute simultaneity—and the global foliation that corresponds to it—is the central conceptual innovation of relativity. Letting a physically important foliation back in, even via definition from the quantum state, looks like step backward.

The argument in favor of counting such a theory as fundamentally relativistic turns on how the foliation is determined. If the rule can be implemented using only the standard relativistic metric and the quantum state, then one can argue that the theory as a whole remains relativistic, even though part of the dynamics adverts to the derived foliation.

Finally, one can adopt a straightforward rejection of the theory of relativity as the final, complete account of space-time structure and just posit a foliation in terms of which the guidance equation is formulated. No one would argue that such a theory is fundamentally relativistic: It maintains that the theory of relativity missed out on an essential feature of space-time. One does not reject any of the usual relativistic structure—light-cones, proper time, and so forth—but supplements it with an additional piece of intrinsic structure. (That is, somewhat ironically, this position holds that the Einsteinian description of the structure of space-time is not complete. It is not that the relativistic metric does not represent real space-time structure, it is just that it does not represent all the objective space-time structure.) Einstein's theory would have missed this structure because it was not tasked with accounting for the physical phenomena, such as violations of Bell's Inequality, that depend on it. From this perspective, it is not surprising that the theory of relativity does not provide a complete accounting of space-time structure, since it was not designed to reproduce characteristically quantum-mechanical effects.

John Bell considered this sort of resolution to the tension between relativity and quantum theory but was unenthusiastic. In his paper "Beables for Quantum Field Theory," he sketches a way to formulate a QFT and finally confronts the question of "serious Lorentz invariance" (i.e., whether the theory is fundamentally relativistic). After some attempts to clarify exactly what the condition of serious Lorentz invariance is, he concludes:

> So I am unable to prove, or even formulate clearly, the proposition that a sharp formulation of quantum field theory, such as that set out here, must disrespect serious Lorentz invariance. But it seems to me that this is probably so.
>
> As with relativity before Einstein, there is then a preferred frame in the formulation of the theory . . . but it is experimentally indistinguishable. It seems an eccentric way to make a world.[4]

It is not entirely clear how to elevate the term "eccentric" into a valid criticism. The general drift seems to be that if a foliation of space-time plays a fundamental dynamical role in a theory, then one would expect some experiments to be able to reveal the foliation empirically. But the empirical predictions of quantum theory—even in Bohmian mechanics, where the foliation plays a central dynamical role—are insensitive to the exact foliation. It remains hidden from empirical view, even as it plays an ineliminable physical role in the theory.

[4] Bell (2004), Chapter 19, p. 180.

How objectionable is this difficulty? One might contend that space-time, by itself, is not observable: It has no intrinsic sensible qualities. At best, we infer the structure of space-time from the observable characteristics of matter in space-time. Why expect that all space-time structure should be empirically discoverable? It is, one might say, surprising that any of the fundamental structure is accessible experimentally and would be surprising if all of it were.

A competing argument is that it requires some sort of "conspiracy" or fine-tuning to shield a fundamental physical structure from empirical observation. Why would Nature make the structure and make use of it, just to hide it from view?

Neither of these rhetorical ploys on its own should carry much weight. The fine-tuning/conspiracy criticism requires more specific detail about the theory under investigation. Given the basic structure of the laws (e.g., the guidance equation), do the constants of nature have to be specially picked to render the foliation empirically inaccessible? Do the laws themselves look somehow ad hoc or contrived or artificial? In the case of the nonrelativistic guidance equation, none of these characterizations is applicable. It is, if anything, a remarkably simple and natural equation. Yet the preferred foliation in the nonrelativistic theory is empirically hidden. That is, although the exact outcome of an experiment may depend on the absolute time order of the distant laboratory operations, one cannot determine from the observable outcomes what that time order is.[5] The mere fact that some fundamental physical structure is empirically inaccessible (according to the theory) is not in itself proof of anything conspiratorial.

If a theory postulates a structure that is empirically inaccessible, then one can rightly wonder whether a different theory, shorn of that structure, can produce the same empirical predictions. But that question, too, can only be investigated in the individual case. The role of the preferred foliation in the formulation of the pilot wave theory is so deeply embedded in the dynamics

[5]The way the outcome depends on the time order of distant events turns on the exact initial conditions of the experiment, and those cannot be empirically determined without interfering with the experiment itself.

that there is no obvious way to do without it. Hence the main options for such a theory: Derive a foliation from something else you already accept or postulate a foliation as an additional piece of space-time structure.

Nonrelativistic quantum theories make use of absolute simultaneity in different ways. It is meaningless to ask what it takes to adapt "quantum mechanics" to a relativistic context, since different exact theories are structured differently. It appears that collapse theories—with both flash and matter density ontologies—can be formulated using only relativistic space-time structure and no preferred foliation of the space-time. This can be accomplished in part because these theories are fundamentally stochastic, with irreducibly probabilistic dynamics. Due to this indeterminism, these theories do not endorse specific counterfactuals about what would have happened under different circumstances, such as if one experimental arrangement had been used rather than another. In contrast, the nonrelativistic pilot wave approach is deterministic and so does support definite counterfactual claims. This seems to require reference to a foliation, which may be either derived or fundamental. There is no obvious bar to extending a Many Worlds approach to a relativistic setting, since it requires neither any collapse nor any guidance equation for the local beables. But that is in part because it is not obvious what local beables the Many Worlds theorist postulates in the first place. The seeming immunity of Many Worlds to these problems may be due to its obscurity on that issue of fundamental importance.

Local Beables for QFT

Particles are fundamentally different from fields. A classical point particle has a trajectory: a collection of space-time points that it occupies, which forms a continuous curve in the space-time. The motion of the particle is that trajectory. Other properties of the particle (its mass, charge, etc.) are attributed to it in connection with dynamical laws with the aim of making precise predictions about the trajectory.

In contrast, a classical field has no trajectory. A field is defined everywhere in space-time, having a value at every point. It may be possible that a field take a "zero value" at a point, but that need not be understood as the absence of the field altogether. Furthermore, classical fields, such as the electromagnetic field, typically take nonzero values almost everywhere. Fields, unlike particles, do not differ in where they are but rather in what values they have at each location.

These differences in the nature of classical fields and particles yield further conceptual differences. Particles can be counted: One can ask how many distinct particle trajectories pass through a given region of space-time. There is some exact number of classical electrons in a given region. But there is only one electromagnetic field in the region, no matter what its state may happen to be.

Even though in this sense classical particles and fields are fundamentally different sorts of things, it is easy to see that each could give rise to the same macroscopic behavior. A physical body of water is fundamentally particle-like due to the molecular structure of water. Look on a fine enough scale, and the water is not spread out continuously in space. But at a gross macroscopic scale, a field-like description of the water (such as given in fluid mechanics) works perfectly well. Ocean currents can be usefully mathematically modeled by continuous functions giving the height of the water, even though at fine-enough scales, these functions do not represent the physical situation. Conversely, it is easy to see how a fundamentally field-like object could exhibit particle-like behavior. Coherent, nondispersing waves in a medium can have trajectories that approximate particle trajectories so long as one does not look too closely. If the field values are small almost everywhere save in some continuous thin regions of space-time, those regions can act like particles, be effectively countable, and so forth.

Since we have been pursuing Bell's strategy of connecting the ontology of an exact fundamental physical theory with the empirical content of the theory via the local beables, we must confront the question of local beables for relativistic QFT. It is tempting to just read the answer off the name. Clearly, it seems, the theory is a theory of fields, so any local ontology must be an ontology of fields. It might be further remarked that one can arrive at the mathematics of the theory by quantizing classical field theory. What "quantization" means, in the simplest terms, is taking a classical theory that postulates certain physical magnitudes (e.g., the value of a field or the position of a particle) and "putting the hats on": The classical quantity is replaced in the new theory by an *operator* that operates on a quantum state. There are rules (or at least guidelines) about how these operators should relate to one another algebraically. In this way, one arrives at nonrelativistic quantum mechanics by replacing the position and momentum quantities that characterize classical particles with position and momentum operators. One also then needs a collection of wavefunctions (or density matrices) for these operators to operate on. In the same mathematical vein, one can take the relativistic theory of the electromagnetic field, which attributes a field value to each point in space-time, and replace these quantities with field operators that are associated with space-time locations. Having thus started with a classical field theory as a mathematical template, it is plausible to assume that any local beables in the resulting theory must themselves be fields.

There is another reason to assume that the local ontology of relativistic QFT should be fieldlike rather than particlelike: the observed phenomenon of particle creation and annihilation. For example, an experiment may be naively described in following way. A high-energy proton circulating clockwise around the Large Hadron Collider strikes a high-energy proton circulating counterclockwise, creating an explosion of new particles—electrons, muons, positrons, protons—that fly out in all directions with different energies. Although this description is couched in terms of particles, the collection of particles described at the end of the experiment is quite different from the pair of particles at the beginning. If we think in terms of particles, they do not appear to exist eternally but come into and go out of existence.

In the nonrelativistic pilot wave theory we have discussed—Bohmian mechanics—this cannot happen. We developed that theory using the configuration space for a fixed number of particles as the space in which the wavefunction is defined (Dürr et al. 2006). Given such a wavefunction, the guidance equation then determines how the particles move, but in this motion, the number of particles is always conserved. In contrast, QFT both allows the number of particles in a situation to change and admits of quantum states of indefinite particle number. An operator in QFT, called the "number operator," is employed in the quantum recipe to make predictions about how many particles of a certain kind one will "find" in a particular experimental situation. As a concrete example, one might create a bubble chamber, in which trails of condensed liquid drops appear, seemingly much like the contrail that follows a jet plane. Charged particles such as electrons create these trails, and so by counting trails, one can count electrons. But certain quantum states in the theory make only probabilistic predictions about this number—a particular collision might create three electrons or four, and the number of observable tracks can be accordingly variable. The collision creates a quantum state of indefinite electron number.

This possibility makes it tempting to deny the existence of any fundamental particles at all. If particles exist, the thought goes, there must at any given time be a definite, exact number of them determined by the number of distinct trajectories. But in a state of "indefinite particle number," no such exact number exists, so there can't be any particles at all. Instead there is a field that can, in particular circumstances, act in a more-or-less particle-like way. The very same field can, at different times, have different numbers of distinct localized lumps in it (corresponding to a definite particle number), but it can also be in a more amorphous state that does not admit of a "particle interpretation."

Given all the considerations in favor of a field-like ontology of local beables for relativistic QFT, it may come as a surprise to read John Bell's paper "Beables for quantum field theory."[6] Bell asks what sort of local beables would be reasonable to postulate in an exact version of QFT. He is largely guided by mathematical tractability in his investigation. The first suggestion he considers

[6] Bell (2004), Chapter 19.

has a field-like character: energy density. A classical field, defined continuously over space-time, produces a continuous distribution of energy. It is easy to see how such a continuous local beable at microscopic scale could relate to the observed outcomes of experiments: Where we say there is a "particle track" through a cloud chamber, the exact theory postulates a localized increase in the energy density.

But introducing such an energy density as a local beable faces technical mathematical problems. The operators associated with energy density in different locations fail to commute, which stymies attempts to attribute definite local energy densities to both locations. And the whole point of the local beables is that they exist and have values regardless of whether they are being "measured." Energy densities might have played that role, but they cannot easily be introduced into the theory for purely mathematical reasons.

Bell suggests that we "fall back then on a second choice—fermion number density. The distribution of fermion number in the world certainly includes the positions of instruments, instrument pointers, ink on paper. . . . and much much more."[7] Fermions (spin-1/2 particles, e.g., electrons and quarks) are prototypical particles, and making the fermion number density in every space-time region a beable implies that a definite physical fact exists about how many fermions are there. So in moving from energy density to fermion number density, Bell switches from a field-like local beable to a particle-like local beable.

Bell's comment about the positions of pointers, instruments, and ink reinforces the point made above. Field-like and particle-like (and flash-like) local beables at microscopic scale are all equally capable of playing the methodological role required of them: serving as the physical basis for observable macroscopic conditions. What is perhaps more surprising is that the particulate local ontology of fermion number density can play this role in the context of QFT, despite the considerations brought above in favor of the local beable being some sort of field. What, for

[7] Bell (2004), p. 175.

example, of the observation that there are quantum states of indefinite particle number?

Bell's approach to QFT is a pilot wave approach: The role of the quantum state of a system is to help define a dynamics for the local beables, the fermions. So the existance of quantum states that are not eigenstates of the number operator is of no more significance than the fact that most quantum states in nonrelativistic quantum mechanics are not eigenstates of the position operator: The Bohmian particles in the pilot wave theory nonetheless always have definite exact positions. Similarly, in Bell's theory, there is always a definite number of fermions at each space-time location. Further, the total number of fermions can change. The terms "particle creation" and "particle annihilation" are taken literally.

Technically, this means that the configuration space for the system is not merely the set of all configurations of N particles, but rather the set of all configurations of all possible numbers of particles (including zero). The wavefunction is defined as a function on this larger, more complicated configuration space. The dynamics of the fermions becomes indeterministic, with the analog of the guidance equation yielding probabilities for various different evolutions of fermion number density. (Bell's own theory is defined over a discrete space-time, like a lattice, rather than a continuous space-time.) Given an initial quantum state and an initial distribution of fermions, the probabilities for later fermion configurations—including configurations with more or fewer fermions—can be derived.

In sum, QFTs can be formulated to solve the measurement problem just as a nonrelativistic quantum theory does: via the postulation of local beables whose behavior underwrites the empirical predictions of the theory. The same variety of microscopic local beables—including flashes, particles, and a continuous matter density—are available to serve this purpose. Detailed theories postulating each of these are constructed in different ways. As we have seen, one can construct theories using flashes and matter densities that employ only the relativistic metrical structure of space-time, while particle theories (with variable particle number) seem to require a preferred foliation. That foliation, in turn, could either be derived from something else or postulated as an additional piece of space-time structure.

Textbook presentations of relativistic QFT mention none of this. They proceed along the lines of textbook presentations of quantum mechanics. Certain operators are identified as the "observables" of the theory and the Quantum Recipe is used to calculate probabilistic predictions for the outcomes of "measurements" of these quantities. No account is given of how to determine, by physical analysis, which observable a concrete experiment measures (if any); nor is any physical account of the measurement interaction itself on offer. In QFT the observables are indexed by space-time locations, just as field values are indexed by space-time locations, and it is sometimes assumed that such an "observable" can be "measured" by experimental conditions set up only in that location. But because no beables are specified, we are as at much of a loss about how to understand the exact ontology of standard textbook QFT as we are with standard textbook quantum mechanics.

The mathematics of QFT is imposing, so much of the philosophical literature on it is challenging for the nonexpert. Two introductions written for philosophers are listed in the Further Reading section for this chapter.

We have barely scratched the surface of QFT, and an introductory book such as this is not the place to go into detail, much less press on to the speculative heights of quantum gravity or string theory. But no matter how complex these theories become, they remain quantum theories, or rather Quantum Recipes: mathematical predictive apparatuses in need of physical theories to underpin and account for their empirical effectiveness. Our basic questions—Is the wavefunction informationally complete? Does the quantum state collapse? What are the local beables?—arise for them in just the same way, and the options available are basically the same. No matter how sophisticated the mathematics gets, or how adept one gets at using it, one still does not understand any of these theories until one understands how they address these most fundamental questions. Unlike the mathematics of nonrelativistic quantum theory (which must be replaced by fancier gadgets as one moves on to field theory), the fundamental physical questions we have chewed over remain. So long as they remain unanswered, the basic quest for understanding physical reality is unfinished. And so long as physicists ignore or dismiss these questions, that basic quest has not yet begun.

Further Reading

Two useful introductory works are a paper by Huggett (2000) and the book by Teller (1995).

References

Albert, David. *Quantum Mechanics and Experience.* Cambridge, MA: Harvard University Press, 1992.

Albert, David. *After Physics.* Cambridge, MA: Harvard University Press, 2014.

Albert, David, and Barry Loewer. "The Tails of Schrödinger's Cat." In *Perspectives on Quantum Reality*, edited by Robert Clifton. Dordrecht: Kluwer, 1996, pp. 81–92.

Barrett, Jeffrey. *The Quantum Mechanics of Minds and Worlds.* Oxford: Oxford University Press, 1999.

Bassi, Angelo, and Hendrik Ulbricht. "Collapse Models: From Theoretical Foundations to Experimental Verifications." *Journal of Physics*: Conference Series 504 (2016): 012023.

Becker, Adam. *What Is Real?* New York: Basic Books, 2018.

Bedingham, Daniel, Detlef Dürr, GianCarlo Ghirardi, Sheldon Goldstein, and Nino Zanghì. "Matter Density and Relativistic Models of Wave Function Collapse." *Journal of Statistical Physics* 154 (2014): 623–31.

Bell, John Stewart. *Speakable and Unspeakable in Quantum Mechanics*, second edition. Cambridge: Cambridge University Press, 2004.

Beller, Mara. *Quantum Dialogue: The Making of a Revolution.* Chicago: University of Chicago Press, 1999.

Bohm, David. "A Suggested Interpretation of Quantum Theory in Terms of 'Hidden' Variables." *Physical Review* 85 (1952): 166–92.

Born, Max. *The Born-Einstein Letters*, translated by I. Born. New York: Walker, 1971.

Bricmont, Jean. *Making Sense of Quantum Mechanics.* Cham, Switzerland: Springer International, 2016.

Cushing, James. *Quantum Mechanics: Historical Contingency and the Copenhagen Hegemony.* Chicago: University of Chicago Press, 1994.

Cushing, James, Arthur Fine, and Sheldon Goldstein, eds. *Bohmian Mechanics and Quantum Theory: An Appraisal.* Dordrecht: Kluwer, 1996.

Deutsch, David. "Comment on Lockwood." *British Journal for the Philosophy of Science* 47 (1996): 222–28.

Deutsch, David. "Quantum Theory of Probability and Decisions." *Proceedings of the Royal Society of London* A4 55 (1999): 3129–37.

Dürr, Detlef, and Stephan Teufel. *Bohmian Mechanics: The Physics and Mathematics of Quantum Theory.* Dordrecht: Springer, 2009.

Dürr, Detlef, Sheldon Goldstein, and Nino Zanghì. "Quantum Equilibrium and the Origin of Absolute Uncertainty." *Journal of Statistical Physics* 67 (1992): 843–907.

Dürr, Detlef, Sheldon Goldstein, James Taylor, Roderich Tumulka, and Nino Zanghì. "Topological Factors Derived from Bohmian Mechanics." *Annales de l'Institut Henri Poincaré* A 7 (2006): 791–807.

Dürr, Detlef, Sheldon Goldstein, and Nino Zangì. *Quantum Physics without Quantum Philosophy.* Dordrecht: Springer, 2013.

Dürr, Detlef, Sheldon Goldstein, Travis Norsen, Ward Struyve, and Nino Zanghì. "Can Bohmian Mechanics Be Made Relativistic?" *Proceedings of the Royal Society* A 470 (2014): 20130699.

Einstein, Albert. *Out of My Later Years.* New York: Philosophical Library, 1950.

Einstein, Albert, Boris Podolsky, and Nathaniel Rosen. "Can Quantum-Mechanical Description of Physical Reality Be Considered Complete?" *Physical Review* 47 (1935): 777–80.

Everett, Hugh. *The Everett Interpretation of Quantum Mechanics*, edited by Jeffrey Barrett and Peter Byrne. Princeton, NJ: Princeton University Press, 2012.

Feynman, Richard, Robert Leighton, and Matthew Sands. *The Feynman Lectures on Physics.* Reading, MA: Addison-Wesley, 1975.

Fraser, Doreen. "Quantum Field Theory: Underdetermination, Inconsistency and Idealization." *Philosophy of Science* 76 (2009): 536–67.

Fuchs, Christopher. "Quantum Bayesianism at the Perimeter." *Physics in Canada* 66, no. 2 (2010): 77–81.

Ghirardi, GianCarlo. *Sneaking a Look at God's Cards.* Princeton, NJ: Princeton University Press, 2005.

Ghirardi, GianCarlo, A. Rimini, and T. Weber. "Unified Dynamics for Microscopic and Macroscopic Systems." *Physical Review* D 34 (1986): 470–491.

Ghirardi, GianCarlo, R. Grassi, and F. Benatti. "Describing the Macroscopic World—Closing the Circle within the Dynamical Reduction Program." *Foundations of Physics* 25 (1995): 5–38.

Greaves, Hilary. "Understanding Deutsch's Probability in a Deterministic Multiverse." *Studies in the History and Philosophy of Modern Physics* 35 (2004): 423–56.

Harrigan, Nicholas, and Robert Spekkens. "Einstein, Incompleteness and the Epistemic View of Quantum States." *Foundations of Physics* 40 (2010): 125–57.

Huggett, Nick. "Philosophical Foundations of Quantum Field Theory." *British Journal for the Philosophy of Science* 51 (2000): 617–37.

Landau, L. D., and E. M. Lifshitz. *Quantum Mechanics: Non-Relativistic Theory*, second edition. Bristol: J. W. Arrowsmith, 1965.

Maudlin, Tim. *Quantum Non-Locality and Relativity.* Malden, MA: Wiley-Blackwell, 2011.

Maudlin, Tim, *Philosophy of Physics: Space and Time*. Princeton, NJ: Princeton University Press, 2012.

Ney, Alyssa, and David Albert. *The Wavefunction: Essays in the Metaphysics of Quantum Mechanics.* Oxford: Oxford University Press, 2013.

Norsen, Travis. "The Pilot-Wave Perspective on Quantum Scattering and Tunneling." *American Journal of Physics* 81 (2013): 258–66.

Norsen, Travis. *Foundations of Quantum Mechanics: An Exploration of the Physical Meaning of Quantum Theory.* Cham, Switzerland: Springer International, 2017.

Okon, Elias, and Daniel Sudarsky. "Benefits of Objective Collapse Models for Cosmology and Quantum Gravity." *Foundations of Physics* 44 (2014): 114–43.

Penrose, Roger. *The Emperor's New Mind: Concerning Computers, Minds and The Laws of Physics.* Oxford: Oxford University Press, 1989.

Perle, Philip. "Combining Stochastic Dynamical State-Vector Reduction with Spontaneous Localization." *Physical Review* A 39 (1989): 2277–89.

Pusey, Matthew, Jonathan Barrett, and Terry Rudolph. "On the Reality of the Quantum State." *Nature Physics* 8 (2012): 475–78.

Saunders, Simon, Jonathan Barrett, Adrian Kent, and David Wallace (eds). *Many Worlds? Everett, Quantum Theory and Reality.* Oxford: Oxford University Press, 2010.

Schrödinger, Erwin. "Discussion of Probability Relations between Separated Systems." *Mathematical Proceedings of the Cambridge Philosphical Society* 31, no. 4 (1935): 555–63.

Shankar, Ramamurti. *Principles of Quantum Mechanics.* New York: Plenum Press, 1980.

Teller, Paul. *An Interpretive Introduction to Quantum Field Theory.* Princeton, NJ: Princeton University Press, 1995.

Tumulka, Roderich. "A Relativistic Version of the Ghirardi-Rimini-Weber Model." *Journal of Statistical Physics* 125 (2006): 821–40.

Wallace, David. "In Defense of Naivete: The Conceptual Status of Lagrangian QFT." *Synthese* 151 (2006): 33–80.

Wallace, David. "Taking Particle Physics Seriously: A Critique of the Algebraic Approach to Quantum Field Theory." *Studies in the History and Philosophy of Modern Physics* 42 (2011): 116–25.

Wallace, David. *The Emergent Multiverse: Quantum Theory According to the Everett Interpretation.* Oxford: Oxford University Press, 2012.

Wallace, David, and Christopher Timpson. "Quantum Mechanics on Spacetime I: Spacetime State Realism." *British Journal for the Philosophy of Science* 61 (2010): 697–727.

Index